## Comments from Accomplished Peop<br>*Security for Wireless Sensor Ne1*<br>*Identity-Based Cryptogra<sub>p</sub>...,*

*Security for Wireless Sensor Networks using Identity-Based Cryptography* provides a step-by-step approach to everything you'll need to know about wireless sensor network security. The book also has information about many topics relevant to wireless sensor network security planning, design, and implementation of Identity-Based Cryptography, which is a superior security solution with better resistance against known attacks.

*—John R. Vacca*
Author, Professional Writer, Editor, Reviewer and IT Consultant

Getting the right balance of systems security is a hard enough task in traditional networking, let alone securing a wireless network composed of sensors with high resource limitations that, nonetheless, can collaborate to perform complex tasks. This book serves as an excellent starting point for practitioners in the industry and students in academia to come up to speed with security issues in wireless sensor networks.

*—Vijay K. Gurbani, Ph.D.*
Distinguished Member of Technical Staff, Bell Laboratories, Alcatel-Lucent
Adjunct Professor, Department of Computer Science, Illinois Institute of Technology

Strong identity mechanisms are a hard sell for today's Net because people feel that they know who they're talking to. In the Internet of Things with its focus on unfeeling machine-to-machine communications, we must replace "feeling secure" with actually being secure. Wireless sensors are the eyes, ears, noses and cat-whiskers of the Internet of Things. Yet these sensors are connected by a network that can be subverted by hostile parties, risking that their output could be diverted or false information inserted. Given the vast number of sensors to be deployed, cryptographic identity systems may be the only manageable approach to trusting our senses.

*—Dean Willis*
Chair, IETF SIP Working Group, 1999-2009

# Security for Wireless Sensor Networks using Identity-Based Cryptography

# Security for Wireless Sensor Networks using Identity-Based Cryptography

Harsh Kupwade Patil

Stephen A. Szygenda

CRC Press
Taylor & Francis Group
6000 Broken Sound Parkway NW, Suite 300
Boca Raton, FL 33487-2742

First issued in paperback 2016

© 2013 by Taylor & Francis Group, LLC
CRC Press is an imprint of Taylor & Francis Group, an Informa business

No claim to original U.S. Government works

Version Date: 20121009

ISBN 13: 978-1-138-19953-8 (pbk)
ISBN 13: 978-1-4398-6901-7 (hbk)

**Visit the Taylor & Francis Web site at**
**http://www.taylorandfrancis.com**

**and the CRC Press Web site at**
**http://www.crcpress.com**

# Contents

# Dedication

In loving memory of
Koki and Tiger

# Foreword

We apologize to Dr. Rehana Yasmin, Dr. Eike Ritter, and Dr. Guilin Wang for failing to properly cite their work, which appears throughout Chapter 5.

The science of cryptography provides the essential foundation for secure communications. We use cryptography everywhere in our everyday communications: our Skype calls are encrypted; our e-commerce web sites are authenticated and encrypted by SSL; and our smartphone apps are digitally signed. But cryptography has always presented difficulties for implementation.

The basic problem in traditional symmetric key cryptography has always been how to share secret keys. In order to establish a secure communication channel, the sender and recipient have to agree on a secret key in a secure way, which poses a chicken-and-egg dilemma.

On the surface, public key cryptography seemed to solve the key sharing problem because one's private key is not shared and the public key can be distributed to everyone. However, public key cryptography created new challenges for an infrastructure of trusted certificate authorities to distribute and manage digital certificates.

Identity-based cryptography (IBC) was a groundbreaking idea to generate public keys from one's public identity, eliminating the need for public key certificates. Messages can be encrypted and sent by deriving the encryption key from the recipient's identity. This novel approach has led to many interesting possibilities that researchers have been exploring.

This book is the fruition of several years of research by my former colleagues at Southern Methodist University to investigate IBC for wireless sensor networks. Wireless sensor networks have a number of difficult security issues related to secure routing, authentication, key distribution, and intrusion detection. This work is a bold pioneering attempt to apply IBC to solve these major security problems. The latter part of the book takes a particularly interesting cross-layer approach to security across different protocol layers.

I believe this book will be a valuable reference on IBC theory and practice that researchers and practitioners will want on their bookshelves for many years. In addition, it will certainly stimulate new directions in research in wireless network security. It was my pleasure to have been involved in the early stages of this work, and I hope you will enjoy reading the book.

*Thomas M. Chen*
Swansea University
Editor-in-Chief, IEEE Network, 2009–2011

# Preface

The use of cryptography in wireless sensor networks (WSNs) is an emerging technology that has dramatically advanced in recent years through the efforts of numerous researchers and practitioners. Chapter 1 of this book introduces WSNs. A WSN network typically consists of low-cost, low-powered, and multifunctional sensor nodes that are deployed in a region of interest. These nodes are small in size and are equipped with sensors, embedded microprocessors, and radio transceivers. Hence, they not only have sensing capacity but also data processing and communication capabilities. They communicate over short distances through the wireless medium and collaborate to accomplish a common task. Such networks have a wide range of military and civilian applications. Due to their widespread applications in numerous areas, WSNs have received an enormous amount of attention from both academia and industry.

Once deployed, sensor nodes can autonomously organize themselves into a wireless communication network. Distinguished from traditional wireless networks, WSNs are characterized by denser levels of node deployment; greater unreliability of sensor nodes; and severe power, computation, memory, and bandwidth constraints. As a result, security in WSNs has been an ever-increasing challenge in such resource-constraint networks. Security mechanisms in such networks are essential to protect data integrity and confidentiality, authentication, access control, user privacy, and continuity of service.

The unique characteristics and constraints on WSNs present many new challenges to their development and application. In Chapter 2, we look at the legacy attacks that were inherited from the traditional wireless networks, as well as new attacks that are adapted to WSNs.

The science of cryptography addresses the problems associated with providing efficient tools to secure data or information. A number of different approaches have been developed over the years. Historically, cryptography was dominated by symmetric key cryptography and mainly dealt with methods to transmit information in a confidential manner. During the late 1970s, asymmetric cryptography emerged from the pioneering work by Diffie and Hellman. Their findings were published in a paper titled "New directions in cryptography" (Diffie and Hellman 1976). However, they were unable to come up with a concrete mathematical proof of their proposed scheme. In 1978, Rivest, Shamir, and Adleman came up with a convincing algorithm that conceptualized public key encryption and called it the RSA algorithm (Rivest et al. 1978). In 1984, Shamir introduced identity-based cryptography (IBC) (Shamir 1984), which forms the basis for this book. He constructed an identity-based signature (IBS) scheme using the existing RSA function (Rivest et al. 1978) but was unsuccessful in constructing an identity-based encryption (IBE) scheme, which remained a long-standing problem for almost a decade. In 2001, Boneh and Franklin came up with an independent solution using the concept of bilinear maps (Boneh and Franklin 2001). This led to a new era of research in IBC. Since then, IBC has become a fascinating area of research in modern cryptography.

Before dealing specifically with IBC, in Chapter 3 we present some of the theory of elliptic curves, assuming that the reader has a modest background in elementary number theory and in groups and field theory. Subsequently, the basic idea of IBE is introduced.

Chapter 4 attempts to cover all important IBE schemes and its applicability in WSNs.

We review broadcast authentication and encryption schemes in WSNs in Chapter 5.

In Chapter 6, we review the interesting problem of key distribution in sensor networks and explore some interesting key-distribution schemes that are based on IBC, which forms the basis for secure routing in WSN (Chapter 7).

Adaptive cross-layer security plays a crucial role in the security of modern WSNs. Chapter 8 emphasizes the need for cross-layer security across all layers of the WSN protocol stack. In addition, it provides examples for cross-layer security using IBC. Chapter 9 reviews some implementation results of IBC on existing sensor node platforms.

This book is intended to serve as a reference and to provide a concise view of IBC in WSNs. It offers in-depth coverage of identity-based cryptographic security theory, technology, and practice as they relate to established technologies, as well as recent advancements in the field. Furthermore, it explores practical security solutions to WSNs using IBC. The purpose of this book is to provide a comprehensive and systematic introduction to the fundamental concepts, major issues, and effective security solutions in wireless sensor networking.

The primary audience for this book consists of researchers and practitioners in industry and academia, as well as security technologists and engineers working on, or interested in, security for WSNs.

The authors are deeply indebted to the myriad of researchers in numerous fields, including mathematics, cryptography, communications, computers, and security, who have provided the foundations for this exciting field of research. It is the confluence of many ideas, from various fields of research and many researchers, that allows us to inch ahead in the technology marathon in which we have chosen to participate.

# Acknowledgments

I am indebted to Dr. Stephen A. Szygenda for his support, guidance, and encouragement in the writing of this book. I would like to thank Dr. Thomas M. Chen for his continued support throughout my career. I would also like to thank Dr. Diego F. Aranha for all the fruitful conversations on the topic of implementing identity-based cryptographic algorithms on sensor nodes.

I would like to thank Richard O'Hanley at CRC Press for giving us the opportunity to write this book. I would also like to thank Theron Shreve and Marje Pollack for providing several rounds of comments on each chapter regarding matters of style, organization, and coverage.

Finally, the book would have never been completed without the encouragement of my parents; my brother, Kunal; and my wife, Akshaya.

# About the Authors

## Harsh Kupwade Patil

Dr. Harsh Kupwade Patil is a Technical Advisor in the Dallas office of Fish & Richardson P.C. His practice supports patent prosecution in a broad range of technologies, particularly in network and system security and applied cryptography. Prior to joining the firm, Dr. Kupwade Patil was involved in collaborative research work with Cisco Systems Inc., Bell Laboratories–Alcatel-Lucent, and Sipera Systems Inc. (acquired by Avaya Inc.). He holds a Master's Degree in Electrical Engineering and a Ph.D. in Applied Science from Southern Methodist University.

## Stephen A. Szygenda

Professor Szygenda is the former Dean of the School of Engineering at Southern Methodist University and The University of Alabama–Birmingham. He has held the position of Chairman of the Electrical and Computer Engineering Department at The University of Texas–Austin, where he also held the Clint Murchison Jr. Chair of Free Enterprise and was the founding Director of the Texas Center for Technology Development and Transfer. During his academic endeavors, Dr. Szygenda acquired extensive government and industry research funding, received numerous awards, consulted for more than 50 international companies and universities, served on the board of directors of a number of companies, and published more than 200 papers. He received his Ph.D. from Northwestern University.

In industry, Dr. Szygenda has served as President of SBI Inc., COMSAT General Integrated Systems, the Rubicon Group, and Comprehensive Computing Systems and Services Inc. He was also a member of the technical staff at Bell Telephone Laboratories.

Dr. Szygenda is a pioneer in the areas of Simulation, CAD, Fault Tolerant Computing, Telecommunications, Software Engineering, Entrepreneurship, Strategic Planning, Technology Transfer, Business Management, and Economic Development.

# Chapter 1

# Introduction to Wireless Sensor Networks

## In This Chapter

- An introduction to wireless sensor networks
- History of WSN
- WSN architecture and protocol stack
- Mobile ad-hoc networks versus WSN
- Challenges in WSN

## 1.1 Introduction

Recent advances in sensing technology, wireless communication, and digital computing techniques have led to the creation and subsequent proliferation of wireless sensor networks (WSN). A WSN can be broadly described as a network of nodes that makes a collaborative effort in sensing certain specified data around its periphery and thereby controls the surrounding environment (Verdone et al. 2008). It is usually a concatenation of a data acquisition network and a data dissemination network. The data acquisition network consists of the actual sensor nodes along with the mobile or stationary base station (sink node); the data dissemination network is a collection of wired and wireless networks that is involved in postprocessing of the acquired data. However, the acquisition network is severely constrained in computing, storage, and power levels when compared to the dissemination network. A stark distinction between WSN and other legacy wireless networks is the decentralized architecture. In addition, the higher scalability ratio in connection with node deployment and hazardous operating environments distinguish WSN from legacy wireless networks such as cellular or mobile ad-hoc networks (MANET).

## 1.1.1 WSN Applications

The past two decades have seen an enormous amount of increase in WSN applications (Akyildiz et al. 2002). Applications can be broadly categorized into military and civilian types. Military applications may include remote sensing, battlefield monitoring, object protection, and intelligent guiding of un-manned moving objects. Civilian applications may include environmental monitoring such as habitat monitoring, disaster monitoring, and air and water pollution control. In addition, WSN may also be used in health care, acoustic and video surveillance, industrial process control, and home intelligence.

### 1.1.1.1 Civilian Applications

Sensors can be used to detect physical parameters such as temperature, pressure, humidity, sound, light, etc. They can also be used to detect air and water quality, and composition of earth. Furthermore, attributes of an object such as height, weight, position, and speed can be used as sensing parameters.

WSN significantly reduce deployment costs because they can be applied in hostile environments such as battlefields, ocean depths, and outer space. Among the earliest applications of sensor networks are environmental monitoring, which further includes habitat monitoring, air and water quality monitoring, hazard monitoring, and disaster monitoring.

- *Habitat monitoring:* The living conditions of wild animals and plants can be monitored using remote sensors. Moreover, rare species of wild animals/plants can be under constant observation using WSN (Naumowicz et al. 2010).
- *Air and water quality management:* Sensors can be placed on the tip of a volcano to detect any eruption, or placed in the middle of the ocean on its floor to detect the possibility of tsunami, or they can be used for ocean pollution prevention (Werner-Allen et al. 2005; Khan and Jenkins 2008).
- *Hazard monitoring:* Sensors can be used to detect any biological or chemical hazard—for example, detecting a chemical or biological weapon. In addition, they can used to detect any leakage of toxic chemical waste or gases in a chemical process plant (Chen et al. 2010).
- *Disaster monitoring:* Natural disasters such as forest fires and flash floods can be easily detected by deploying WSN in such volatile areas. Seismic sensors can be planted to detect the direction and magnitude of an earthquake. They can also be deployed around nuclear plants, oil and gas pipelines, and strategic bridges.
- *Health care:* WSN have widespread applications in health care, in that they can be used to monitor patients with disabilities and thereby reduce the need for

health care personnel (Jafari et al. 2005). They can also be used in the study of the behavior of patients by remotely monitoring and helping behavioral scientists in their research. Such networks can be used for early detection of clinical deterioration of patients in hospitals or can provide emergency care in large disasters such as earthquakes or flooding (Ko et al. 2010).

- *Smart homes:* Wireless sensors have played a major role in improving "smart home" technologies (Byoung-Kug et al. 2008). For example, a smart refrigerator equipped with embedded WSN technology can sense if the home owner is low on groceries and can send a text message/email to the owner's cell phone with an alert message. WSN have played a crucial role in enhancing safety measures for cars by sensing the possibility of a collision and automatically applying brakes (Lai et al. 2010). In addition, wireless sensors can be used in automating utility bills by remotely reading utility meters installed in homes to measure usage of gas, water, or electricity.

### 1.1.1.2 Military Applications

WSN have become an integral part of many military applications (Lee et al. 2009). The ease of deployment of WSN in any ad-hoc environment and the ability to self-configure have made them the number-one choice in battlefield environments. Moreover, the use of such intelligent and autonomous networks reduces casualties and takes the art of warfare to a whole new level. For example, motion-detecting sensors can be used to track the presence of enemy forces on land, air, or water and can facilitate constant surveillance.

Sensors can be placed on manned or unmanned robots to guide them around obstacles and help them interact better among themselves. In addition, they can be used on missiles for pinpoint-precision attacks.

## 1.2 History of WSN

Work in WSN began as early as the onset of the Cold War, with the development of a Sound Surveillance System (SOSUS) by the United States to detect and monitor the Soviet Union's submarines. Over the years, many sophisticated acoustic sensor networks have been developed for submarine surveillance. In the early 1980s, the U.S. Defense Advanced Research Project Agency (DARPA) initiated research in WSN by initiating the Distributed Sensor Network (DSN) program (Chee-Yee and Kumar 2003). Basically, DSN consisted of many spatially distributed sensing nodes that operated autonomously and made independent routing decisions. During this time, the U.S. Advanced Research Projects Agency Network (ARPANET) was successful in establishing about 200 hosts at various research institutes and universities, and the founders of ARPANET were curious to know whether the approaches used in designing communication protocols for ARAPNET could be used in DSN. Most of the pivotal technological constituents for a DSN were identified at the Distributed

Sensor Nets workshop in 1978 (Proceedings of the Distributed Sensor Nets Workshop. Pittsburgh 1978). The key components included acoustic sensors, a focus on developing new self-location algorithms for distributed networking, and development of new protocols for digital communication among sensors. Since research in artificial intelligence (AI) was also supported by DARPA, the workshop included talks on the use of AI in signal processing. Furthermore, it focused on various distributed problem-solving techniques. During the same time, researchers at Carnegie Mellon University (CMU) developed an operating system kernel called Accent (Rashid and Robertson 1981). Accent provided transparent and flexible access to distributed resources, which was a necessity for a fault-tolerant DSN. Later, Accent evolved into one of the most popular operating systems, known as Mach (Rashid et al. 1989). A practical example of DSN was a helicopter tracking system developed at the Massachusetts Institute of Technology (MIT) (Myers et al. 1984). This system used a distributed array of acoustic microphones along with knowledge-based signal-processing techniques to track helicopters. However, researchers soon realized that tracking multiple targets in a distributed environment was a greater challenge than a centralized architecture. Although researchers in the 1980s and early 1990s had a vision of the future of WSN, the miniaturization technology was not there to support their goals. Sensors at that time were quite large, which presented an impediment to practical implementation.

The 1980s saw the emergence of multiple-hypothesis tracking algorithms that were decomposed for distributed implementations (Chong et al. 1990). These algorithms were basically designed to target strenuous situations involving high target density and false alarms. Although radar and national power grid networks have existed for decades, the term "wireless sensor network" came into vogue in the late 1990s with the advent of microelectromechanical systems (MEMS) (Pierret 1990; Senturia 2001). MEMS have been the driving technology in manufacturing tiny, low-cost, and low-powered sensor nodes. This new paradigm of research in WSN has attracted a lot of attention and has paved the way to a new era of network research in a highly dynamic and ad-hoc environment. It has also broadened the scope of various civilian and military WSN applications. For example, body sensor networks, vehicular sensor networks, and many other such applications that require tiny sensors with a limited sensing range and processing power have emanated from the introduction of MEMS. In this new era of WSN research, DARPA initiated a new research program called SensIT, which facilitated a new environment for new algorithms in an ad-hoc, decentralized networking paradigm (Kumar and Shepherd 2001). During this time, the IEEE saw the potential of WSN in diverse fields and established the IEEE 802.1 WPAN Task Group, which produced *IEEE Standard 802.15.4–2003: Wireless Medium Access Control (MAC) and Physical Layer (PHY) Specifications*. This standard defines the characteristics of the data-link and physical layer for low-data-rate wireless personal area networks. The advantages of such networks include short-range communication with higher probability of reliable data transmission and extreme low cost in deployment with minimal human supervision. Based on IEEE Std. 802.15.4, a group of companies formed an alliance called the ZigBee Alliance (www.zigbee.org) to produce standards for low-power wireless networking. This new technology is gaining widespread acceptance in many different industrial and governmental organizations. ZigBee reuses the data-link and physical layer specified by the IEEE Std. 802.15.4 and introduces the network layer for different topologies. It also provides a framework for application-layer development.

## 1.3 WSN Architecture

A sensor network typically consists of a collection of low-cost, low-powered sensor nodes distributed in an ad-hoc fashion. Each of these networks has one or more sink nodes or base stations, which is usually a resource-abundant computer with sufficient storage, computing, and communication capabilities (Figure 1.1). These sink nodes send queries or commands to their respective sensor nodes, which then collaborate with each other to accomplish the given task. In addition, the sink nodes act as a gateway between the sensor network and the outside network (Internet or cellular network).

A typical sensor node is equipped with the actual sensing equipment for data aggregation, embedded microcontroller for instantaneous processing of data, and a radio transceiver for digital communication. Usually, nodes communicate over a short distance and use a wireless medium for all communications. Each sensor node senses data from its surrounding environment and sends this captured data to its sink node. Generally, the data or control packets from the sending node travel in a hop-by-hop fashion through every intermediate sensor node to reach the sink node (Figure 1.2).

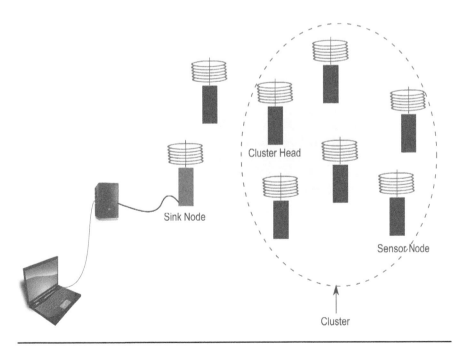

**Figure 1.1.** Wireless Sensor Network

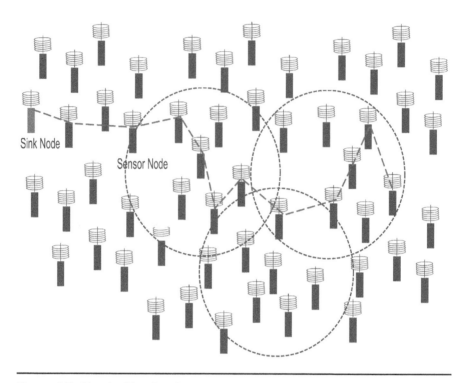

**Figure 1.2.** Hop-by-Hop Routing

However, in certain situations, it is inefficient for each node to forward the captured raw data directly to the sink node, as this could lead to early exhaustion of valuable resources (battery, memory, and computation ability) of each intermediate routing node. Hence, to overcome this issue, the network is further divided into clusters and each cluster has a cluster head. Each cluster head is responsible for locally aggregating data from its cluster nodes and then forwarding it to its sink node. Furthermore, cluster heads can preprocess the raw data before sending it to the sink node.

Depending on the way data is aggregated by sensor nodes, WSN architectures can be classified into homogenous, heterogeneous, and hybrid sensor networks (Nakayama et al. 2007).

### 1.3.1 Homogenous Sensor Networks

In a homogenous network architecture, the cluster heads and the ordinary sensor nodes have the same computational, storage, and communication capabilities as a sink node. In this architecture, the network topology determines the method of data aggregation. Usually, flat and hierarchical architectures (Figure 1.3 and Figure 1.4) are used in a homogenous architecture (Rajagopalan and Varshney 2006).

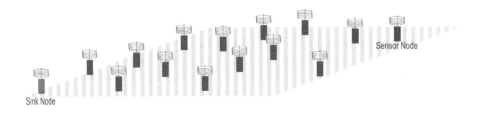

**Figure 1.3.** Flat Architecture in WSN

### 1.3.1.1 Flat Architecture

In a flat architecture, every entity of the network (sink node and sensor nodes) has the same computational and communication capabilities. Hence, it can also be perceived as a peer-to-peer architecture. However, with the increase in the scalability of the sensor nodes, it is rather challenging to have a global addressing scheme. For this reason, a data-centric routing scheme is sought as a pragmatic solution to the increasing scalability issue. In this scheme, the sink node broadcasts a query or command to all the nodes in the network, and only the concerned nodes that have a corresponding match to the query respond to the sink node.

### 1.3.1.2 Hierarchical Architecture

In contrast to the flat architecture, the hierarchical architecture segregates the network into clusters and increases the overall energy efficiency of the network (Shah et al. 2003). In this architecture, the nodes operate in close proximity to their respective

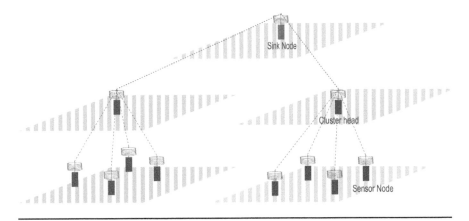

**Figure 1.4.** Hierarchical Architecture in WSN

cluster heads. Hence, nodes with lower energy levels simply capture the required raw data and forward it to their respective cluster heads. Usually, the cluster heads possess more processing and storage capacity than any ordinary sensor node.

In the case of a highly dynamic and a decentralized environment, nodes with higher energy levels can take responsibility for a cluster head and nodes with lower energy levels can be the cluster nodes. Furthermore, adopting this architecture will not only increase the energy efficiency of the network, it will also balance traffic load with an increase in size of the network. In certain situations, data aggregation can be performed at the cluster heads instead of the sink node to improve the efficiency of the entire network.

### 1.3.2 Heterogeneous Sensor Networks

In a heterogeneous sensor network, the sink node or base station can be mobile (Shah et al. 2003; Chatzigiannakis et al. 2006). Hence, the mobile sink node can move randomly to any of the sensing regions and collect data by interacting closely with the sensor nodes. In addition, the sink's mobility will increase the overall energy level of the system.

## 1.4 Architecture of Sensor Nodes

A typical sensor node consists of a sensing unit, a processing unit, and a transceiver (Figure 1.5). The sensing unit consists of actual sensors, and it is the only interface to the outside environment.

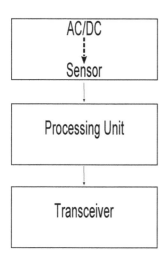

**Figure 1.5.** Architecture of Sensor Node

The processor unit contains the microprocessors which process information received from the sensing unit. Due to energy constraints, the processing unit performs partial processing of data and leaves the extensive processing to the sink node. Hence the processors operate in different modes, so that they can lower their activity in situations of lower battery power. However, the decision to change to a different processing mode requires collaboration with other layers. (Layered architecture is discussed in Section 1.5).

Operating frequency and operating voltage determine the amount of energy required by the microprocessor. New voltage scheduling schemes have been adopted to save energy (dynamic voltage scheduling), which take operating voltage and frequency into consideration during processing (Pillai and Shin 2001). Recent research has shown that the amount of energy required by the communication unit is much greater than that of the processing unit.

The communication unit consists of a short-range transceiver, customized to operate in a resource-constrained environment. The limited battery energy forces the transceiver to adopt a hop-by-hop mechanism over the end-to-end mechanism adopted by legacy IP networks. For this reason, sensor nodes are placed in close proximity to each other, so that the node discovery phase and routing process consume less battery power.

## 1.5 The Protocol Stack in WSN

WSN operate in a resource-constrained environment and therefore deviate from the traditional Open System Interconnection (OSI) model. A WSN stack is usually comprised of six layers: an application layer, middleware, transport, network, data link, and a physical layer. In addition to these six layers that are mapped to each sensor node, there are three more planes that span across the entire sensor network and have more visibility to address issues such as mobility, power, and task management (Figure 1.6).

- *Application layer:* The application layer aims to create an abstraction of the main functions of the sensing application, thereby making the lower software and hardware levels transparent to the end user. The application layer involves several processes running simultaneously, and handles user requests relating to data aggregation, location finding, sleep/awake cycle control, time synchronization, authentication, encryption, key distribution, and other security measures. It also defines the order and format of message exchange between the two communication parties.
- *Middleware:* The middle layer provides an Application Program Interface (API) for applications existing in the upper layers, and may involve complex functionalities such as resource sharing and task management.
- *Transport layer:* The transport layer is responsible for flow and congestion control. It also performs error control to detect corrupted frames that arrive from lower layers. Due to the severe operating environment and lesser transmission power,

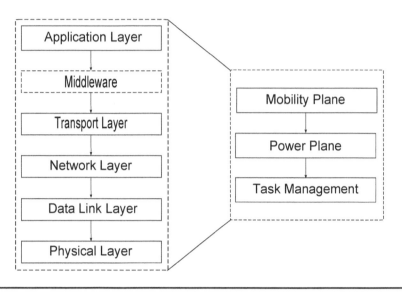

**Figure 1.6.** WSN Protocol Stack

it is difficult to achieve high end-to-end link reliability compared to traditional wireless networks. In addition, the transport layer performs fragmentation of sender data and reassembly of received data frames.

- *Network layer:* The network layer's primary goals are to perform routing operations and self-configuration. It is responsible for link failures and provides regular updates to neighboring nodes. However, assuring network connectivity at all times is a major challenge due to dynamically changing network topology. The routing protocols in WSN are very different from traditional routing protocols because of the need to optimize network life by performing intelligent routing.
- *Data link layer:* The data link layer is an interface between the network and physical layer. It is further subdivided into two modules: Medium Access Control (MAC) and Logical Link Control (LLC). The MAC module plays a critical role in conserving network life by efficiently allocating medium access to the contending nodes. The LLC is on top of the MAC layer and is responsible for cyclic redundancy check (CRC), sequencing information and adding appropriate source and destination information.
- *Physical layer:* The physical layer is responsible for converting digital bits into analog symbols and vice versa. It involves modulation and demodulation, frequency selection, power control, and symbol synchronization. WSN usually operate in frequencies ranging from 915 MHz to 2.4 GHz. It is recommended to use a lower-frequency band, as there is higher attenuation when operating in higher-frequency bands. However, with the limited availability of the bandwidth in the lower frequencies, WSN is forced to operate at higher frequencies. The environment in which sensors are operating plays a major role in signal attenuation. Thus,

sensors placed on the ground or floating on water experience greater attenuation and consequently require higher transmit power. The choice of modulation scheme is one of the prime factors in deciding the transmit power. The modulation scheme decides the bit error rate (BER), spectrum efficiency, and number of bits per symbol. For example, an M-ary modulation scheme is able to transmit more bits per symbol than other binary modulation schemes such as Phase Shift Keying (PSK). However, M-ary schemes result in higher BER and require more transmit power than the binary modulation schemes. Hence binary modulation schemes are more applicable to WSN.

- *Mobility plane:* Sensor nodes can be fixed on moving objects such as animals, vehicles, people, etc; which will lead to a dynamic topology. In the event of some mobility by sensor nodes, the mobility in collaboration with the network layer is responsible for maintaining the list of active neighboring nodes. It is also responsible for interacting periodically with other neighboring nodes' mobility planes, so that it can create and maintain a table of active, power-efficient routes.
- *Power plane:* The power plane focuses on the awareness of power at each horizontal and vertical layer. It is responsible for shutting off the sensors if they are not participating in any routing decisions or simply if the sensing activity is complete. Power planes of each node work collectively on deciding efficient routes to sink nodes and maintain the sleep/awake cycles of sensor nodes.
- *Task management plane:* The task management plane is responsible for achieving a common goal by taking properties of each layer and across each layer in a power-aware manner.

## 1.6 MANET Versus WSN

A mobile ad-hoc network (MANET) is a network of mobile nodes communicating through a wireless medium and with an ability to self-configure itself (Tavli and Heinzelman 2005). The ability of each node to move freely in any direction results in a topology that is very dynamic. MANET has been studied for decades and is closely related to WSN. However, recent research work has shown that WSN can be categorized as a separate research topic. Let us look at the similarities and dissimilarities of MANET and WSN.

### 1.6.1 Similarities Between MANET and WSN

MANET and WSN are both wireless ad-hoc networks with distributed architectures. The nodes in both types of network are powered by batteries. Intermediate nodes in each of the networks may be involved in routing decisions. One of the major drawbacks to both types of network is that they operate in an unlicensed spectrum, resulting in radio interference from other radio technologies. Self-management is a key feature in both types of network.

### 1.6.2 Differences Between MANET and WSN

One of the fundamental differences between MANET and WSN is that MANET nodes (PDA, laptop computers, cell phones, and other mobile radio terminals) are in close association with human beings. However, sensor nodes are in close proximity to the surrounding environment. Consequently, sensor nodes are more scalable than MANET nodes, and can range from a few hundreds to thousands of nodes. Deployment scenarios are as diverse as sensor nodes placed on the edge of an active volcano or in the middle of an ocean bed. Therefore the likelihood of sensor nodes being damaged is higher when compared to MANET nodes. For this reason, the topology of WSN is very dynamic and is not solely dependent on the node mobility, as in MANET. Hence, the topology of the network is also dependent on the sleep/awake cycle. Sensor nodes have ROM memory of a few kilobytes and processing speeds of a few megahertz. On the other hand, MANET nodes have large storage space and high-speed processors that can record and play video files.

## 1.7 Challenges in WSN

In general, WSN pose significant technical challenges to communication, data processing, and security. (Threats to security will be discussed in detail in Chapter 2.) Moreover, harsh and dynamic operating environments, along with energy and bandwidth constraints, have increased challenges to node/network discovery, routing, data aggregation, querying, and network management. Below, we discuss a few of these challenges.

### 1.7.1 Battery Life and Power Consumption

The need for power control to increase energy efficiency has been one of the main challenges in WSN. Usually, sensor nodes are battery-operated but function in perilous environments, making frequent replacement of batteries difficult. It is important for sensor nodes to monitor battery life, as there is a possibility of the node receiving high current that lasts for a long duration, resulting in rapid depletion of the battery. Hence, caution should be exercised by periodically shutting down the node when it is not in the active routing path or during sensing and processing operations. Another technique is to use complex data processing techniques to reduce battery consumption.

The network life is heavily dependent on the battery life, and the need for new energy efficient schemes that take every layer of the sensor node into consideration has been a foremost challenge in WSN. In addition, predicting the necessary power level for transmitting control and data packets is quite complex. The sensor's power level directly affects the transmission range, thus changing the routing path. For example, if a neighboring node is damaged, the transmitting sensor node should transmit at a higher power level, thus reducing its battery life and reducing its chances of participating in any further intermediate routing operations. Lower power levels can lead to interference and cause congestion in the channel and the network. Therefore, power control can be categorized as a cross-layer design problem, as it affects each and every layer of the stack.

## 1.7.2 Power Conservation Mechanism

Power conservation can be divided into active and passive schemes (Figure 1.7). Passive schemes simply turn off the transceiver when there is no sensing, processing, or communication activity; active schemes work toward improving node operations.

Power conservation above the data link layer can be categorized into structured and nonstructured mechanisms. In structured mechanisms, all nodes are organized into a structure, such as a group of clusters. Each cluster contains a head node, and the power conservation mechanism is performed on each head node. It may perform a synchronization of sleep/awake schedules on a group of nodes in a cluster to conserve bandwidth, and assign proxy nodes in situations of head-node failure. SPAN is an example of such a mechanism.

On the contrary, a nonstructured based mechanism allows each sensor node to schedule its sleep/awake cycles independently. A node may consider neighboring nodes' sleep/awake cycles in tuning its own cycle. Beacon/hello messages are used to discover its nearest neighbors. Some examples include TDMA scheduling, energy efficiency using sleep-mode TDMA schedule, S-MAC, and self-stabilizing deterministic TDMA.

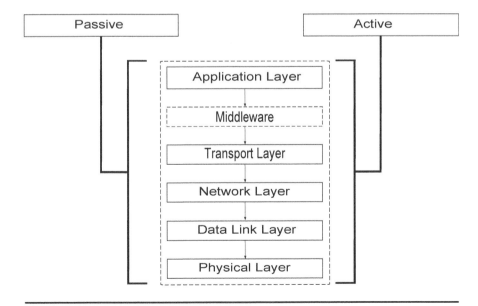

**Figure 1.7.** Power Conservation Mechanism

### 1.7.3 Communication Media

A huge part of energy consumption is dependent on the medium of communication. Since WSN uses radio-frequency (RF) communication, a great deal of effort has gone into modeling the wireless channel so that it can help in characterizing the behavior of WSN. Furthermore, multipath fading adds new complexities to transmission of wireless data in WSN. Research suggests that the radio irregularity model (RIM) is more applicable than disc or spherical propagation models.

### 1.7.4 Integrated Circuit (IC) Technology

The use of custom-designed ICs rather than commercial ICs for WSN has greatly reduced the consumption of battery power and has increased the lifetimes of the sensor nodes. Furthermore, with advances in miniaturization technology, the size of the sensor nodes has been reduced drastically and scalability has increased. However, integration of node supply voltages with custom ICs is still an open research problem (Cook et al. 2006).

## 1.8 The Routing Problem in WSN

Routing in sensor networks has been very complex and challenging due to their intrinsic characteristics that differentiate them from other legacy wireless networks. First and foremost, the assignment of a global addressing scheme is not possible due to the usage of a relatively large number of sensor nodes. Furthermore, the ad-hoc, decentralized nature of WSN compels the nodes to self-organize themselves to form a connected network. Although routing in WSN has been more data-centric as opposed to address-centric, it does raise serious security issues concerning authentication of sensor nodes in a data-centric networking model. Second, the acceptance of a data-centric model requires an efficient data aggregation model (discussed in Section 1.9.3) to reduce redundancy and retransmission of packets. In addition, the resource-constrained nature of WSN requires new energy-aware routing schemes. Some of the challenges include the following.

- Routing protocols in WSN should aim to reduce the energy consumed per packet. However, one of the major challenges is the uneven distribution of sensor nodes. Due to the uneven topology, some intermediate nodes may expend more energy when transmitting packets. In addition, if nodes are along an active routing path, the energy depletion rate will be higher than for other nodes. Therefore, choosing an active routing path should consider the power levels of each node in the network on a time basis.
- Proposing a global addressing scheme in a WSN network is very challenging, due to the relatively large number of sensor nodes. In addition, IP headers can be

very heavy for a processor with limited processing capacity. Hence, connectivity to the IP network is made only through the sink node.

- Designing position-aware sensor networks is another challenge. For example, consider a mobile node in an enemy territory that is able to sense some movement and relay this suspicious activity to the sink node. Soon after the relay, if the node is compromised or destroyed, it could be difficult for the sink node to find the exact location, as it is unable to reach the compromised node. Hence, addressing schemes in WSN should take location into consideration.

## 1.9  Broadcasting and Multicasting

One of the fundamental operations in WSN is data dissemination from sensor nodes to their sink nodes and vice versa. This is usually achieved by performing a broadcast, multicast, or geocast operation.

### 1.9.1  Broadcasting

Broadcast operation is a "one-to-all" operation, in which packets are disseminated from a source to all the nodes in the network. Broadcasting can be broadly divided into simple broadcast schemes, probability-based broadcasting, neighborhood-aware broadcast mechanism, location-aided broadcasting, energy-efficient broadcasting, and reliable broadcasting (Ni et al. 1999).

#### 1.9.1.1  Simple Broadcasting

Simple broadcasting does not require any prior knowledge of the network or the states in which the nodes of the network exist. A few examples of this mechanism include blind broadcast, probability-based broadcast, distance-based broadcast, area-based broadcast, and counter-based broadcast. However, in most of the broadcast schemes, there is a possibility of duplicating packets, which could be devastating in a WSN environment.

#### 1.9.1.2  Multicasting

The multicasting mechanism is "one-to-many." In this operation, packets are disseminated from one source to multiple destinations. For example, in WSN, a sink node can send multicast messages to a selected set of sensor nodes. One of the main objectives of introducing this scheme is to improve the overall efficiency of the system by decreasing the number of relaying nodes. Multicasting in WSN can be broadly divided into tree-based and location-based schemes. Examples of tree-based multicasting mechanisms in WSN include multicast-enabled, ad-hoc, on-demand distance vector routing, centralized power-aware multicasting, and localized power-aware multicasting. Location-based multicasting mechanisms include scalable energy-efficient asynchronous dissemination, geographic multicast routing, and two-tier data dissemination.

### 1.9.1.3 Geocasting

In geocasting, dissemination of packets is done in a set of nodes belonging to a specified area. For example, the area might be a rectangle, square, circle, or a polygon. This scheme assumes prior knowledge of the location coordinates of each node before dissemination of packets. Geocasting can be further divided into nonguaranteed geocasting and guaranteed geocasting, depending on whether the sensor nodes can receive geocast packets or not. Examples of nonguaranteed geocasting mechanisms include unicast routing with area delivery and directed flooding-based geocasting. Guaranteed geocasting mechanisms include simple flooding, geocasting via efficient broadcasting, and geocasting via face routing.

## 1.9.2 Network Discovery

The a-priori information about the location of nodes can play a major role in proper functioning of the network. However, due to the dynamic environment of WSN, it may not be possible to have prior knowledge of the topology. Hence, in such ad-hoc, decentralized architectures, the topology has to be constructed in real time. In addition, periodic updates about the topology should be disseminated to all the nodes in the network, as there could be deletion or addition of new sensors in the network. While dynamic topology discovery in WSN has been an ongoing research topic, self-discovery of each node with reference to its respective sink nodes is an equally important research topic in WSN. Although the Global Positioning System (GPS) has been used extensively in MANET, its direct application in WSN is not feasible due to the resource-constrained environment of WSN.

## 1.9.3 Data Aggregation in WSN

The high-density deployment of sensor nodes in a particular region can lead to a phenomenon called spatial correlation. In this event, identical sensing of data can occur by nodes in close proximity to each other. Hence, there is a need to minimize redundancy in data received from the node's immediate neighbors, before transmitting it to the sink node. It is during such scenarios that data aggregation plays a critical role in reducing redundancy and minimizing retransmissions, and in doing so, increasing the overall energy efficiency of the network. Furthermore, the adoption of a data-centric approach for networking over an address-centric approach has provided more emphasis for data aggregation.

## 1.10  Summary

In this chapter, we introduced WSN and looked into its history. We provided a comparative analysis of WSN and MANET that led to the conclusion that the security solutions applicable to MANET cannot be applied directly to WSN. Furthermore, we reviewed different architectures for WSN and assessed each layer of the WSN stack. We concluded the chapter with an evaluation of the existing challenges to providing security in a resource-constrained environment such as WSN.

## 1.11  References

Akyildiz, I. F., W. Su, Y. Sankarasubramaniam, and E. Cayirci. Wireless sensor networks: A survey. *Computer Networks* 38, no. 4 (March 2002): 393–422.

Byoung-Kug, Kim, Hong Sung-Kwa, Jeong Young-Sik, and Eom Doo-Seop. The study of applying sensor networks to a smart home. *Fourth International Conference on Networked Computing and Advanced Information Management.* IEEE, 2008, 676–681.

Chatzigiannakis, I., A. Kinalis, and S. Nikoletseas. Sink mobility protocols for data collection in wireless sensor networks. *Proceeding of the 4th ACM Workshop on Mobility Management and Wireless Access.* ACM, 2006, 52–59.

Chee-Yee, Chong, and S. P. Kumar. Sensor networks: Evolution, opportunities, and challenges. *Proceedings of the IEEE* 91, no. 8 (2003): 1247–1256.

Chen, Lei, Shuang Yang, and Ying Xi. Based on ZigBee wireless sensor network the monitoring system design for chemical production process toxic and harmful gas. *International Conference on Computer, Mechatronics, Control and Electronic Engineering (CMCE).* IEEE, 2010, 425–428.

Chong, C.Y., S. Mori, and K. C. Chang. Distributed multitarget multisensor tracking. In Y. Bar-Shalom (ed.), *Multitarget-Multisensor Tracking: Applications and Advances.* Artech House, 1990, 247–295.

Cook, B., S. Lanzisera, and K. Pister. SoC issues for RF smart dust. *Proceedings of IEEE* (IEEE) 94, no. 6 (2006): 1177–1196.

*IEEE Std. 802.15.4—2003: Wireless Medium Access Control (MAC) and Physical Layer (PHY) Specifications.* IEEE, 2003.

Jafari, R., A. Encarnacao, A. Zahoory, F. Dabiri, H. Noshadi, and M. Sarrafzadeh. Wireless sensor networks for health monitoring. *Proceedings of 2nd Annual International Conference on Mobile and Ubiquitous Systems: Networking and Services.* IEEE, 2005, 479–481.

Khan, A., and L. Jenkins. Undersea wireless sensor network for ocean pollution prevention. *Proceedings of 3rd International Conference on Communication Systems Software and Middleware and Workshops.* IEEE, 2008, 2–8.

Ko, JeongGil, Chenyang Lu, M. B. Srivastava, J. A. Stankovic, A. Terzis, and M. Welsh. Wireless sensor networks for healthcare. *Proceedings of the IEEE* (IEEE) 98, no. 11 (October 2010): 1947–1960.

Kumar, S., and D. Shepherd. SensIT: Sensor information technology for the warfighter. *Proceedings of 4th International Conference on Information Fusion.* IEEE, 2001, 3–9.

Lai, Xin, Jun Li, Wei Zhong, and Yuan Cao. Fast and accurate WSN positioning in emergency logistics. *Second International Conference on Networks Security Wireless Communications and Trusted Computing (NSWCTC).* IEEE, 2010, 277–280.

Lee, Sang Hyuk, Soobin Lee, Heecheol Song, and Hwang Soo Lee. Wireless sensor network design for tactical military applications: Remote large-scale environments. *Proceedings of the 28th IEEE Conference on Military Communications.* IEEE, 2009, 1–7.

Myers, C., A. Oppenheim, R. Davis, and W. Dove. Knowledge based speech analysis and enhancement. *International Conference on Acoustics, Speech and Signal Processing.* IEEE, 1984, 162–165.

Nakayama, H., N. Ansari, A. Jamalipour, and Y. Nemot. Fault-resilient sensing in wireless sensor networks. *Computer Communications* 30, no. 11 (2007): 2375–2384.

Naumowicz, T., et al. Wireless sensor network for habitat monitoring on Skomer Island. *Proceedings of 35th Conference on Local Computer Networks (LCN).* IEEE, 2010, 882–889.

Ni, S., Y. Tseng, Y. Chen, and J. Sheu. The broadcast storm problem in a mobile ad hoc network. *Proceedings of ACM MOBICOM.* ACM, 1999, 151–162.

Pierret, R. F. *Introduction to Microelectronic Fabrication.* Addison-Wesley, 1990.

Pillai, P., and K. G. Shin. Real-time dynamic voltage scaling for low-power embedded operating systems. *Proceedings of the Eighteenth ACM Symposium on Operating Systems Principles.* ACM, 2001, 89–102.

Proceedings of the Distributed Sensor Nets Workshop. Pittsburgh. Department of Computer Science, Carnegie Mellon University, 1978.

Rajagopalan, R., and X. Varshney. Data-aggregation techniques in sensor networks: A survey. *IEEE Communications and Surveys and Tutorials,* 2006: 48–63.

Rashid, R., and G. Robertson. Accent: A communication oriented network operating system kernel. *Proceedings of 8th Symposium on Operating Systems.* ACM, 1981, 64–75.

Rashid, R., et al. Mach: A system software kernel. *34th Computer Society International Conference (COMPCON).* IEEE, 1989, 176–178.

Senturia, S. D. *Microsystem Design.* Kluwer Academic, 2001.

Shah, R., S. Roy, S. Jain, and W. Brunette. Data mules: Modeling and analysis of a three-tier architecture for sparse sensor networks. *Ad Hoc Networks* (Elsevier) 1, no. 2–3 (2003): 215–233.

Tavli, B., and W. Heinzelman. *Mobile Ad Hoc Networks: Energy-Efficient Real-Time Data Communications.* Springer-Verlag, 2005.

Verdone, R., D. Dardari, G. Mazzini, and A. Conti. *Wireless Sensor and Actuator Networks.* Elsevier UK, 2008.

Werner-Allen, G., J. Johnson, M. Ruiz, J. Lees, and M. Welsh. Monitoring volcanic eruptions with a wireless sensor network. *Proceeedings of the Second European Workshop on Wireless Sensor Networks.* IEEE, 2005, 108–120.

# Chapter 2

# Security in Wireless Sensor Networks

## In This Chapter

- The various kinds of attacks on WSN
- The attack taxonomy for WSN
- Threats to privacy, control, and availability
- Countermeasures to attacks on WSN

## 2.1 Introduction

Wireless sensor networks have been a promising approach to many diverse applications and are playing a major role in the upcoming age of intelligent networking. However, with the increase in popularity of such networks, there is an ever-increasing demand for better security. In many military applications, for instance, sensor networks are involved in mission-critical tasks, and security therefore is a top priority in such situations. In addition, resource, computing, and environmental constraints on WSN (discussed in Chapter 1) pose new challenges in securing such networks. Security in WSN is especially challenging due to the inherent broadcast nature of wireless communication. Any radio equipment can participate, either passively or actively, in any form of radio communication, by tuning the equipment to the same frequency band as that of the communicating nodes.

The demand for WSN is increasing even faster with the introduction of miniaturization technology. Consequently, there has been mass production of sensor nodes with

19

mediocre hardware, which is usually not tamper-resistant. Hence, nodes can easily be compromised by attackers, leading to a completely new set of attacks. In this chapter, we review the different kinds of attacks that can be carried out on WSN and then briefly look into state-of-the-art security in WSN.

## 2.2 Attacks on WSN

Wireless sensor networks are a subclass of wireless networks in general, so most kinds of attacks that can be directed at wireless networks can be directed at WSN. However, due to the additional challenges as discussed in Chapter 1, WSN breed a whole new set of attacks that can be classified into "mote-class attacks" and "laptop-class attacks" (Karlof and Wagner 2003). In a mote-class attack, the attacker takes control of a few compromised nodes and has capacity no greater than that of these ordinary sensor nodes. Hence, he has to launch attacks without depleting his resources (storage, computation, and bandwidth). Moreover, the attacker is restricted in his coverage due to limited transmission power. In contrast, the laptop-class attacker is resource-abundant (high transmitting power, longer battery life, high-speed processors, and highly receptive omnidirectional antennae). Hence, he has greater coverage and diversity in terms of the attacks that he can launch. Figure 2.1 portrays the attack taxonomy for WSN, while Table 2.1 shows the attack classification through a layered approach.

In general, attacks can be divided into active and passive attacks.

- *Passive attack:* In this type of attack, the attacker is able to intercept and monitor data between communicating nodes, but does not tamper or modify packets for fear of raising suspicion of malicious activity among the nodes. For example, in traffic analysis, the attacker may not be able to decode encrypted data, but can find useful information by analyzing headers of packets, their sizes, and the frequency of transmission. In WSN, reconnaissance can also be performed to understand information exchange between communicating nodes, particularly at data aggregation points. Furthermore, routing information can be exploited using traffic analysis.
- *Active attack:* In this type of attack, the attacker actively participates in all forms of communication (control and data) and may modify, delete, reorder, and replay messages or even send spoofed illicit messages to nodes in the network. Some other active attacks include node capturing, tampering with routing information, and resource exhaustion attacks. Peculiar to WSN, the attacker can modify the environment surrounding sensors, which could affect the sensed phenomena.

### 2.2.1 Threats to Privacy

In WSN, threats to privacy can be further classified into reconnaissance and eavesdropping.

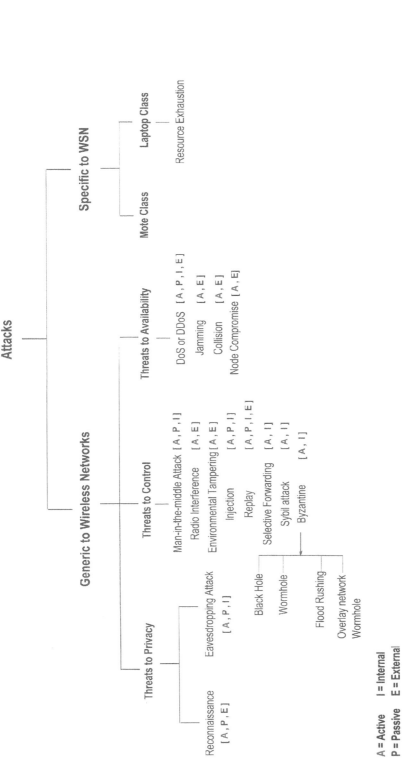

**Figure 2.1.** Attack Taxonomy for WSN

**Table 2.1. Attack Classification Using a Layered Approach**

| Layer | Attack vectors |
|---|---|
| Application | Data corruption and repudiation |
| Transport | Session hijacking and SYN flooding |
| IP layer | Byzantine, resource consumption, location disclosure, wormhole and black hole |
| Data link | Traffic analysis |
| Physical | Interference and jamming |
| Multilayer attack | DoS, replay, man-in-the-middle and replay. |

### 2.2.1.1 Reconnaissance

Reconnaissance refers to intelligent gathering or probing to access the vulnerabilities in a network, to launch a full-scale attack later. Reconnaissance attacks can be further classified into active and passive. Passive reconnaissance attacks include the collection of network information through indirect or direct methods, but without probing the targe; active reconnaissance attacks involve the process of gathering traffic with the intention of eliciting responses from the target.

### 2.2.1.2 Eavesdropping

Eavesdropping is the act of listening secretly to a private conversation. However, in the paradigm of WSN, eavesdropping is an operation to learn the "aggregate data" that is being collected by the entire network. Hence, eavesdropping between two specific sensor nodes may not help the attacker in thoroughly understanding the entire network. It can be further classified into *active* and *passive* eavesdropping.

- *Active eavesdropping:* In this case, the adversary actively sends queries to other nodes in an attempt to instigate them to respond to his queries, and in exchange will be able to comprehend the precise task assigned to the nodes in the network. Usually, the attacker launches a "man-in-the-middle attack" (discussed in the next section) to infiltrate into the network and enforce himself on the active path.
- *Passive eavesdropping:* The attacker inserts himself into the active path, unbeknownst to other nodes in the network. He then passively listens to all traffic sent over the broadcast medium. It may be difficult to detect a passive eavesdropping attack, as the attacker may be operating in stealth mode. (Detection mechanisms are covered in the chapter on intrusion detection systems.)

## 2.2.2 Threats to Control

### 2.2.2.1 Man-in-the-Middle Attack

Man-in-the-middle attack is one of the classical attack vectors that can be executed in a WSN environment (Figure 2.2). In this type of attack, the attacker intrudes into the network and makes an effort to establish an independent connection between a set of nodes and the sink node. The nodes in the network are unaware that the entire flow control is being handled by the attacker. He can be in either a passive or an active state. In a passive state, he simply relays every message among the nodes with the intention of performing an eavesdropping attack. In an active state, he can tamper with the intercepted data in an attempt to break authentication (Authentication is discussed below). The attack can be executed at the physical, data link, network, and application layers (Anand et al. 2005).

### 2.2.2.2 Radio Interference

With the increase in the number of wireless technologies using the same open spectrum band (2.4 GHz, 5 GHz, or 900 MHz), there is bound to be radio interference. For example, in a dense urban environment, where cordless phones share the same spectrum, there is a sharp degradation of individual node performance due to radio interference. Similar problems can be foretold for sensor networks with the increase in sensor nodes per network. The result of such interference could lead to change in the information bits transmitted over the wireless medium, thereby making the bits unintelligible and ultimately being dropped by the receiver (Xu et al. 2008). Hence, radio

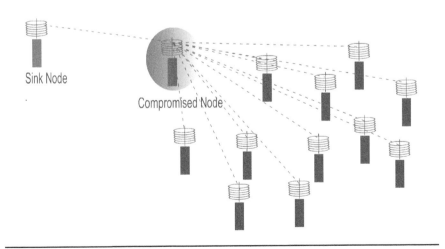

**Figure 2.2.** Man-in-the-Middle Attack in WSN

interference could lead to a denial-of-service attack. The worst-case scenario in radio interference is jamming.

### 2.2.2.3 Injection Attack

After the attacker has clandestinely intruded into the WSN network, he may impersonate a few of the sensor nodes (or even sink nodes) and may inject malicious data into the network. The malicious data might be false advertisement of neighbor-node information to other nodes, leading to impersonation of sink nodes and aggregation of all data.

### 2.2.2.4 Replay Attack

A replay attack is a common attack in WSN, whereby an attacker is able to intercept user data and retransmit user data at a later time. This attack is particularly useful in breaking weak authentication schemes, which do not consider the time stamp when authenticating nodes. This attack is also useful during shared-key distribution processes.

### 2.2.2.5 Byzantine Attack

In a Byzantine attack, the outside adversary is able to take full control of a subset of authenticated nodes that can be used to attack the network from inside. Such attacks by malicious behavior are known as Byzantine attacks. Some examples of Byzantine attacks are black holes, wormholes, flood rushing, and overlay network wormholes.

- *Black hole attacks:* In this type of attack (Figure 2.3), the attacker drops packets selectively, or all control and data packets that are routed through him. Therefore, any packet routed through this intermediate malicious node will suffer from partial or total data loss.
- *Flood rushing attack:* This type of attack (Figure 2.4) is common to wireless networks and exploits the flood duplicate suppression technique. In this attack, the attacker attempts to overthrow the existing routing path by sending a flood of packets through an alternate route, which will result in discarding the legitimate route and adopting the adversarial route. Usual authentication schemes cannot prevent this attack, as the adversaries are authenticated nodes.
- *Wormhole attack:* In this type of attack, two conniving sensor nodes, or laptops, tunnel control and data packets between each other, with the intention of creating a shortcut in the WSN. Such a low-latency tunnel between the two conniving nodes will likely increase the probability of it being selected as an active path. This type of attack is very closely related to the sinkhole attack, because one of the conniving nodes could falsely advertise to be the sink node

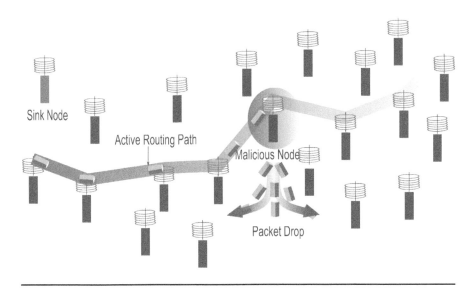

**Figure 2.3.** Black Hole Attack

and thereby attract more traffic than usual. One of the main differences between a Byzantine wormhole and a traditional wormhole is that in a Byzantine wormhole, the tunnel exists between two compromised nodes, while in a traditional wormhole, two legitimate nodes are tricked into believing that a secure tunnel exists between them.

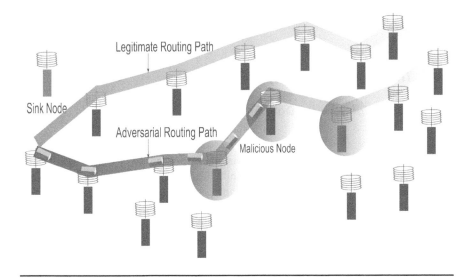

**Figure 2.4.** Flood Rushing Attack

- *Byzantine overlay network wormhole attack:* This type of attack is a variant of wormhole attack, and occurs when the wormhole attack is extended to multiple sensor nodes; resulting in an overlay of compromised nodes. It provides a false illusion, to honest nodes, that they are surrounded by legitimate nodes, resulting in frequent reuse of the adversarial path.

### 2.2.2.6  Sybil Attack

The sybil attack was first introduced by Douceur while studying security in peer-to-peer networks (Douceur 2002), and later Karlof and Wagner showed that this type of attack poses a serious threat to routing mechanisms in WSN (Karlof and Wagner 2003). Sybil is an impersonation attack in which a malicious node masquerades as a set of nodes by claiming false identities, or generating new identities in the worst case (Newsome et al. 2004). Such attacks can be easily executed in a WSN environment because the nodes are invariably deployed in an unstructured and distributed environment, and communicate via radio transmission. They are especially detrimental in applications such as data aggregation, voting systems, reputation evaluation, and geographic routing. Using a sybil attack in location-aware routing, it is possible to be in multiple locations at the same time.

### 2.2.2.7  Sinkhole Attack

In a sinkhole attack, the adversary impersonates a sink node and attracts the whole of traffic to a node or a set of nodes. Similar to a blackhole attack, the attacker takes control of a few compromised nodes and advertises false routing information to its neighbors, thereby luring all traffic to him.

### 2.2.3  Threats to Availability

### 2.2.3.1  Denial of Service (DoS) or DDoS

A denial-of-service attack occurs when an attacker floods the victim with bogus or spoofed packets with the intent to lower the response rate of the victim. In the worst-case scenario, it makes the victim totally unresponsive. For instance, in a WSN environment where nodes have limited computational capacity, a DoS attack from a resource-abundant adversary can overwhelm the nodes by flooding packets, which will exhaust communication bandwidth, memory, and processing power. From an attacker's point of view, this attack is also useful in wireless networks where nodes are required to deliver time-critical data. Jamming (discussed in the next section) the wireless links can also lead to a DoS attack.

An extension of a DoS attack is a distributed DoS attack, where an attacker takes control of a few nodes in the network, leading to a distributed flood attack against the victim.

### 2.2.3.2 HELLO Flood Attack

One of the common techniques to discover neighbors is to send HELLO packets. If a node receives a HELLO packet, it indicates that it is within the range of communication. However, a laptop-class adversary can easily send HELLO packets with sufficient power to convince the sensor nodes that it is in proximity of communication and may be a potential neighbor. The adversary can also impersonate a sink node or a cluster node.

### 2.2.3.3 Jamming

Jamming is one of the most lethal types of attacks in WSN and is a direct way to compromise the entire wireless network. In this type of attack, the attacker jams a spectrum band with a powerful transmitter, and prevents any member of the network in the affected area from transmitting or receiving any packet. Jamming attacks can be divided into constant jamming and sporadic jamming. Sporadic jamming can be very effective at times when a change in one bit of a data frame will force the receiver to drop it. In this kind of attack, it is difficult for the victim to identify whether his band is being jammed intentionally or due to channel interference, and his immediate reaction is usually to increase his transmitting power, thereby depleting resources at a faster rate. Jamming attacks target the physical and MAC layers. Xu et al. discuss four types of jamming attacks (random, reactive, deceptive, and constant) which would result in DoS attacks (Xu et al. 2005). They conclude that detection schemes can be very complex with reference to differentiating malicious attacks from link impairment.

### 2.2.3.4 Collision

Collision attacks target the MAC layer to create costly exponential back-off. Whenever collision occurs, the nodes should retransmit packets affected by collision, thus leading to multiple retransmissions. The amount of energy expended by the attacker is much less than the energy expended (battery exhaustion) by the sensor nodes. Collision attack can be categorized under resource exhaustion attacks.

### 2.2.3.5 Node Compromise

Node compromise is one of the most common and detrimental attacks in WSN. As sensors can be deployed in harsh environments such as a battlefield, ocean bed, or the edge of an active volcano, they are easily susceptible to capture by a foreign agent. In case of a battlefield scenario, the enemy could make an effort to dig into nodes with the intention of extracting useful data (extracting private keys in sensor nodes). Furthermore, it could be reprogrammed and launched into a battlefield to operate on behalf of the enemy.

## 2.2.4 Attacks Specific to WSN

### 2.2.4.1 Attacks on TinyOS Beaconing Protocol

TinyOS Beaconing Protocol uses a breadth-first spanning tree algorithm to broadcast routing updates. The sink node periodically broadcasts updated routing information to its immediate neighboring nodes. These neighboring nodes then rebroadcast this information to their immediate neighbors, and the process continues recursively. During this process, each intermediate node makes a note of its parent node (the parent node is the first node that was able to make contact with its subordinate node and relay the routing information). When all the active nodes are operational, they should send all the sensed data to their parent node. However, this protocol is vulnerable to many attacks. For example, a simple impersonation attack, leading to a sinkhole attack, can totally compromise the entire network (Karlof and Wagner 2003; Sun et al. 2008).

Authentication can be used to prevent such impersonation attacks, but it does not prevent a laptop-class adversary from launching a selective forwarding attack (Figure 2.5), an eavesdropping attack, or a black-hole attack. The attacker creates a wormhole between two conniving laptop-class adversaries. The two laptops are placed near the sink node and the targeted area, respectively. The laptop near the sink node attracts its

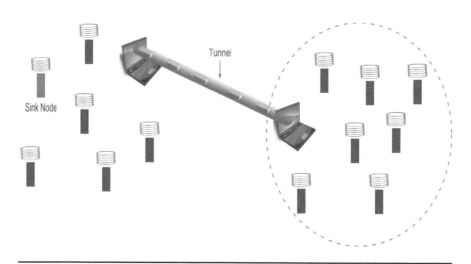

**Figure 2.5.** Selective Forwarding Attack in TinyOS Beaconing

entire neighbor's traffic and simply tunnels these authenticated messages to its colluder. The laptop attacker, close to the sink node, plays a passive role in forwarding these messages. Due to his furtive nature, it is difficult for his neighbors to detect whether he is malicious. Once the authenticated messages reach the remote laptop adversary, he could launch a black-hole attack or a selective forwarding attack.

Let us consider a situation where digital signatures are being used for authentication and, while the routing updates are in progress, the sink node's private key is leaked. (Methods used to leak the private key are beyond the scope of this book.) As soon as the sink node realizes that its private key is being compromised, it immediately broadcasts a new public key. All the nodes in close proximity to the sink node will update their local copy of the sink node's public key. The laptop close to the sink node will perform the same operation and convey this information to its colluding laptop. The remote laptop can now easily impersonate the sink node and launch a sinkhole attack. In addition, he can further create routing loops, which is a resource-exhaustion attack.

### 2.2.4.2 Attacks on Geographic- and Energy-Aware Routing (GEAR)

GEAR (discussed in Chapter 1) proposes a location- and energy-aware, recursive routing algorithm to address the problem of uneven energy consumption in routing in WSN. In GEAR, every node gauges the energy levels of its neighbors along with the distance from the target before making a routing decision. In such situations, a laptop-class attacker can advertise that he has larger energy levels than his neighboring node and attract all traffic to him. Thenceforth, he can execute a sybil, black-hole, or selective forwarding attack.

## 2.3  Countermeasures to Attacks in WSN

As attacks on WSN become more sophisticated, the demand for new security solutions is continually increasing. Hence, an array of new security schemes have been designed and implemented in the past decade (Healyet al. 2009; Perrig et al. 2004). Most of these schemes have been designed to provide solutions on a layer-by-layer basis rather than on a per-attack basis; in doing so, they have left a gap between layers that may lead to cross-layer attacks. We will look more closely at cross-layer security later in this book.

In general, any security suite should ensure authentication, integrity, confidentiality, availability, access control, and nonrepudiation. In addition, physical safety is absolutely necessary to avoid tampering or destruction of nodes. Therefore, construction of tamper-resistant sensor nodes is absolutely necessary. However, such tamper-resistant schemes come at a higher manufacturing cost and are restricted to applications that are not only critical but that use fewer nodes.

- *Authentication:* The main objective of authentication is to prevent impersonation attacks. Hence, authentication can be defined as the process of assuring that the identity of the communicating entity is what it claims to be.

- *Integrity:* The goal of integrity is to affirm that the data received is not altered by an interceptor during communication (by insertion, deletion, or replay of data) and is exactly as it was sent by the authorized sender. Usually, cryptographic methods such as digital signatures and hash values are used to provide data integrity.
- *Confidentiality:* The goal of confidentiality is to protect the data from unauthorized disclosure. A common approach to achieving confidentiality is by encrypting user data.
- *Availability:* The goal of availability is to ensure that the system (network) resources are available and usable by an authorized entity, upon its request. It tries to achieve survivability of the network at all times.
- *Access control:* The goal of access control is to enforce access rights to all resources in its system. It tries to prevent unauthorized use of system and network resources. Access control is closely related to authentication attributes. It plays a major role in preventing leakage of information during a node-compromise attack. One of the conventional approaches to access control is to use threshold cryptography. This approach hides data by splitting it into a number of shares. To retrieve the final data, each share should be received through an authenticated process.
- *Nonrepudiation:* Nonrepudiation can be best explained with an example. Let Alice and Bob be two nodes, who wish to communicate with each other. Let Alice send a message (M) to Bob. Later, Alice claims that she did not send any message to Bob. Hence, the question that arises is how Bob should be protected if Alice denies any involvement in any form of communication with Bob. Nonrepudiation aims to achieve protection against communicating entities that deny that they ever participated in any sort of communication with the victim.

## 2.3.1 Security in WSN Using a Layered Approach

### 2.3.1.1 Security Measures in the Physical Layer

To prevent radio interference or jamming, the two common techniques used are frequency-hopping spread spectrum (FHSS) and direct-sequence spread spectrum (DSSS). In FHSS, the signal is modulated at frequencies such that it hops from one frequency to another in a random fashion at a fixed time interval. The transmitter and the corresponding receiver hop between frequencies using the same pseudo-random code for modulation and demodulation. If an eavesdropper intercepts a FHSS signal, unless he has prior knowledge of the spreading signal code, he will not be able to demodulate the signal. Furthermore, spreading the signal across multiple frequencies will considerably reduce interference.

In DSSS, a spreading code is used to map each data bit in the original signal to multiple bits in the transmitted signal. The pseudo-random code (spreading code) spreads the input data across a wider frequency range compared to the input frequency. In the frequency domain, the output signals appear as noise. Since the pseudo-random code provides a wide bandwidth to the input data, it allows the signal power to drop down below the noise threshold without losing any information. Therefore, this technique is hard for an eavesdropper to detect, due to lower energy levels per frequency and more tolerance to interference.

The above-mentioned schemes can provide security only as long as the hopping pattern or the spreading code is not disclosed to any adversary.

## 2.3.1.2 Security Measures in the Data Link Layer

Link-layer security plays an important role in providing hop-by-hop security. Its protocols are useful in handling fair channel access, neighbor-node discovery, and frame error control. Legacy security protocols such as SSL or IPSec cannot be applied directly to WSN because they do not provide data aggregation or allow in-network processing, which are prime requirements in designing security protocols.

To prevent denial-of-service (DoS) attacks on WSN, it is proposed that each intermediate node in the active routing path perform an authentication and integrity check. However, if a few intermediate nodes in the active path have very low energy levels, and if they are forced to perform authentication checks, they would expend all their energy and disrupt the active path. On the other hand, if we look at end-to-end authentication in WSN, it is more energy-efficient, since the sink node (resource-abundant) is the only node that performs authentication and integrity checks. Nevertheless, this scheme is vulnerable to many types of security attacks (black hole, selective forwarding, and eavesdropping). Hence there is a need for adaptive schemes that consider the energy levels of each node when deciding on the authentication schemes.

Early security approaches focused on symmetric keying techniques, and authentication was achieved using Message Authentication Code (MAC). One of the common MAC schemes is a cipher block chaining message authentication code. However, this scheme is not secure for variable-length input messages. Hence the end user (sensor nodes) have to pad the input messages to be equal to a multiple of the block cipher. Therefore, each node has to waste energy padding input data. To overcome this issue, other block cipher models such as CTR and OCB have been proposed. With reference to confidentiality, symmetric encryption schemes used to protect WSN are DES, AES, RC5, and Skipjack (block ciphers) and RC4 (a stream cipher). Usually, block ciphers are preferred over stream ciphers because they allow authentication and encryption.

A few proposed link-layer security frameworks include TinySec, Sensec, SNEP, MiniSec, SecureSense (Karlof et al. 2004; Perrig et al. 2002), and ZigBee Alliance (www.zigbee.org/Specifications.aspx). However, these schemes have limitations. For example, in Tinysec a single key is manually programmed into all the sensor nodes in the network. A simple node-capture attack on any one of these nodes may result in the leakage of the secret key and compromising of the entire network. A need for a stronger keying mechanism is needed to secure TinySec. In addition, TinySec requires padding for input messages that are less than 8 bytes. It uses block cipher to encrypt messages, and for messages that are less than 8 bytes, the node will have to use extra energy to pad the message before encrypting.

## 2.4 Summary

Although WSN are a subclass of wireless networks, due to the resource-constrained nature of WSN, they breed a whole new set of attacks. In this chapter, we introduced the attack taxonomy for WSN and discussed various kinds of attacks on such networks. Countermeasures to these attacks were discussed from a layered perspective of the WSN stack.

## 2.5 References

Anand, M., Z. G. Ives, and I. Lee. Quantifying eavesdropping vulnerability in sensor networks. Departmental papers, Department of Computer & Information Science, University of Pennsylvania, 2005.

Douceur, J. R. The sybil attack. *First International Workshop on Peer-to-Peer Systems (IPTPS '02)*. 2002.

Healy, M., T. Newe, and E. Lewis. Security for wireless sensor networks: A review. *SAS 2009— IEEE Sensors Applications Symposium*. IEEE, 2009, 80–85.

Karlof, C., and D. Wagner. Secure routing in wireless sensor networks: Attacks and countermeasures. *First International Workshop on Sensor Network Protocols and Applications*. IEEE, 2003, 113–127.

Karlof, C., N. Sastry, and D. Wagner. TinySec: A link layer security architecture for wireless sensor networks. *Second ACM Conference on Embedded Networked Sensor Systems*. ACM, 2004, 162–175.

Newsome, J., E. Shi, D. Song, and A. Perrig. The sybil attack in sensor networks: Analysis & defenses. *Third International Symposium on Information Processing in Sensor Networks, IPSN*. IEEE, 2004, 259–268.

Perrig, A., R. Szewczyk, J. Tygar, V. Wen, and D. Culler. SPINS: Security protocols for sensors. *Wireless Networks* 8, no. 5 (2002): 521–534.

Perrig, A., J. Stankovic, and D. Wagner. Security in wireless sensor networks. *Communications of the ACM* 47, no. 6 (June 2004): 53–57.

Sun, Z., X.-G. Zhang, H. Li, and A. Li. The application of TinyOS beaconing WSN routing protocol in mine safety monitoring. *International Conference on Mechtronic and Embedded Systems and Applications, MESA*. IEEE, 2008, 415–419.

Xu, W., W. Trappe, Y. Zhang, and T Wood. *Proceedings of the Sixth ACM International Symposium on Mobile Ad Hoc Networking and Computing (MobiHoc '05)*. ACM, 2005, 48-57.

Xu, W., W. Trappe, and Y. Zhang. Defending wireless sensor networks from radio interference through channel adaptation. *ACM Transactions on Sensor Networks*, 4, no. 4 (August 2008).

# Chapter 3

# Elliptic Curve Cryptography

## In This Chapter

- Basic concepts of field theory
- Why elliptic curve cryptography
- Pairing-based cryptography
- Weil and Tate pairing

## 3.1 Introduction

Modern cryptography is one of the building blocks in the design of wireless sensor networks (WSN). The two fundamental categories are symmetric key schemes and asymmetric key schemes. In symmetric key schemes, the communicating pair has to agree on secret and authentic keying material before initiating encrypted communication. Hence, a major drawback with such schemes is the requirement of an authenticated and confidential channel for distributing keying material. The need for a constant trusted third party weakens the applicability of symmetric key cryptography in distributed networks. Furthermore, problems with key management (Chapter 6) and nonrepudiation services in symmetric key schemes exacerbate its usage in an ad-hoc network such as WSN. In 1975, Whitfield Diffie and Martin Hellman introduced public-key cryptographic schemes that required only an authentic key exchange between the communicating parties (Diffie and Hellman 1976). This was the first discrete log system. In this scheme, each communicating entity had a private and a corresponding public-key pair such that retrieving the private key from the corresponding public key was computationally infeasible. This revolutionary approach laid the foundation for modern cryptography. In 1977, the first fully conceptualized public-key cryptographic algorithm was proposed by Ron Rivest, Adi Shamir, and Leonard Adleman (RSA algorithm; Rivest et al. 1978). The security of such public-key cryptographic schemes depended on the integer factorization problem or the discrete log problem. In 1984, ElGamal proposed public-key based signature and encryption schemes based on the discrete log problem (ElGamal 1984).

Most security products adopt the RSA algorithm to generate digital signatures and encryption schemes. However, in recent years elliptic curve cryptography (ECC) has gained a lot of attention, as it provides the same level of security as RSA but for a far smaller key size, thus leading to lower processing overhead. Hence, ECC is an excellent substitute for RSA in systems/networks that have constrained resources. This chapter focuses on ECC and, in particular, on paring-based cryptography, which is useful in understanding identity-based cryptography and its application to WSN. The basis for these disciplines in cryptography uses mathematical concepts taken from the field of modern algebra. For completeness, we briefly review the basic building blocks of modern algebra, which will be used in the mathematical derivation of elliptic curve cryptographic techniques to be presented.

## 3.2  Introduction to Modern Algebra

The fundamental elements of modern algebra are groups, rings, and fields. In this branch of mathematics, we are particularly interested in performing algebraic operations on the elements of the defined set, resulting in elements that already belong to the same set. We will look into the specific rules applied to these operations that define the nature of the set. In particular, we are interested in modular arithmetic operations such as addition and multiplication.

### 3.2.1  Groups

Let $(G,*)$ denote a group in which $G$ is a set such that the binary operation $* : G \times G \rightarrow G$ satisfies the following properties:

(i)  Closure:  $\forall\, a, b \in G$

$$a * b \in G \tag{3.1}$$

(ii)  Associative:  $\forall\, a, b, c \in G$

$$a * (b * c) = (a * b) * c \tag{3.2}$$

(iii)  Identity element:  $\forall\, a \in G$,  there exists an element $e \in G$ such that

$$a * e = e * a = a \tag{3.3}$$

(iv) Inverse element: $\forall\, a \in G$, there exists an element $a' \in G$ such that

$$a * a' = a' * a = e \tag{3.4}$$

(v) In addition, a group $G$ is *abelian* if it satisfies the following property:

Commutative: $\forall\, a,b \in G$,

$$a * b = b * a \tag{3.5}$$

If the identity element in the group is 0 and the inverse of $a$ is $-a$, then the group is denoted as an additive group. Similarly, if the identity element in the group is 1 and the inverse of $a$ is $a^{-1}$, the group is denoted as a multiplicative group.

If $|G|$ is finite, then the group $(G,*)$ is finite. Furthermore, the *order* of a group is defined as the number of elements in the finite group.

If $a \in G$, then let $k$ be the least positive integer such that $a^k = 1$, provided $k$ exists. Then the *order* of $a$ [ord($a$)] is defined to be $k$. If $k$ does not exist, the *order* of $a$ is defined to be infinity, $\infty$.

An example of a group is a set of integers $(Z)$ with an additive operation. Let $Z_n$ denote $Z/nZ$ (integer modulo $n$). Meanwhile, not all elements of the set $Z_n$ under multiplication modulo $n$ have multiplicative inverses. Hence, set $Z_n$ under multiplication modulo $n$ is not a group. However, $Z_n^*$ (multiplicative group of $Z_n$) $= \{a \in Z_n \,|\, gcd(a,n) = 1\}$. In particular, if $n$ is a prime, then the set of elements in $Z_n^*$ includes the range $1 \le a \le n-1$.

*Cyclic group:*

A group $G$ is *cyclic* if every element in it can be expressed as a power of a fixed element. For example, if $a \in G$ is a fixed element, then every element in $G$ can be expressed in the form $a^k$, where $k \in Z$ (integers). Hence, $a$ is the generator of the group $G$. Note that every subgroup in a cyclic group is cyclic.

*Lagrange's theorem:*

If $H$ is a subgroup of $G$, then the order of $H$ ($|H|$) divides the order of $G$ ($|G|$). Consequently, if $a \in G$, then ord($a$) divides $|G|$, provided ord($a$) exists.

### 3.2.2 Rings

A ring $(R,+,\times)$ is a set $R$ with binary operations, namely, addition $(+)$ and multiplication $(\times)$, on $R$ and satisfying the following properties:

(i) Closure under multiplication
$\forall\, a,b \in R$

$$a \times b \in R \tag{3.6}$$

(ii)  $(R,+,\times)$ is an abelian group with respect to addition and has 0 as its identity.

(iii)  The operation $\times$ is associative.
$$\forall\, a,b,c \in R,$$

$$a\times(b\times c)=(a\times b)\times c \tag{3.7}$$

(iv)  The multiplicative operation $(\times)$ is distributive under addition $(+)$.

(v)
$$\forall\, a,b,c \in R$$

$$a\times(b+c)=(a\times b)+(a\times c) \tag{3.8}$$

and

$$(b+c)\times a=(b\times a)+(c\times a) \tag{3.9}$$

Example for $R$: The set $Z$ under addition and multiplication is a commutative ring. In addition, the set $Z_n$ under addition modulo $n$ and multiplication modulo $n$ is a commutative ring.

## 3.2.3  Fields

A *field*, denoted $(F,+,\times)$, is a commutative ring in which all nonzero elements of the set have multiplicative inverses. The order of the finite field is defined to be the number of elements in the field. If $F_q$ is a finite field where $q = p^m$ ($p$ = prime number and $m$ = positive number), then $p$ is defined to be the *characteristic* of a field.

The *characteristic* of a field is the least positive integer $m$ such that $\sum_{i=1}^{m} 1$ equals 0. However, it is 0 if

$$\overbrace{1+1+1+1\cdots+1}^{m \text{ times}} \neq 0 \quad \text{ for any } m \geq 1 \tag{3.10}$$

For example, the set $Z$ under the operations of addition and multiplication is not a field, as the only nonzero integers with multiplicative inverses are 1 and $-1$. However, real numbers $(R)$, complex numbers $(C)$, and the rational numbers $(Q)$ form a field.

## 3.3  Prime Fields

Let $F_p$ be a field where $p$ is a prime number. Then integer modulo $p$ is an integer set with elements $\{0,1,2,\ldots,p\text{-}1\}$.

For example, $F_{19}$ is a field with elements $\{0,1,2,\ldots,18\}$, and addition, subtraction, multiplication, and inversion operations can be performed. Table 3.1 shows addition modulo 19, and Table 3.2 shows multiplication modulo 19.

## 3.4 Binary Fields

Binary fields or characteristic-2 finite fields (denoted as $F_{(2^m)}$) are fields of order $2^m$. Conventionally, polynomial basis representation is used to represent $F_{2^m}$. The elements of the binary fields are the binary polynomials with at most degree $m-1$.

$$F_{2^m} = \{a_{m-1}x^{m-1} + a_{m-2}x^{m-2} + \cdots + a_3 x^3 + a_2 x^2 + a_1 x^1 + a_0\} = \sum_{i=0}^{n-1} a_i x^i \quad (3.11)$$

where $a_0 \in \{0,1\}$

Table 3.3 shows that binary field $F_{2^3}$ has 8 binary polynomials with degree 2.

A binary polynomial $f(z)$ of degree $m$ is an irreducible polynomial if it cannot be factored as a product of binary polynomials with each of degree less than $m$. In such cases, multiplication of the field elements is performed using the *reduction polynomial* $f(z)$.

## 3.5 Elliptic Curve Cryptography

One of the fundamental ideas of any public-key cryptographic system is the requirement of a hard problem and that, from such problems, secure public key exchange can be facilitated (Diffie and Hellman 1976). Hard problems fall under the *computational security* model, which measures the computational effort involved in defeating a system by the currently best known method. A proposed technique is supposed to be computationally secure if the perceived level of computation required to defeat the method (using the best known attack) exceeds the computational resources of the hypothesized adversary. In addition, an encryption or digital signature can depend on the same hard problem. In the past 30 years, many hard problems have been proposed, but only two of them have gained a lot of public acceptance in cryptography. These are the integer factorization problem and the discrete log problem over a finite field (Simmons 1999). For instance, the RSA algorithm, named for its inventors, Rivest, Shamir, and Adleman, is based on the difficulty of factoring large numbers. However, RSA is operationally expensive, with a minimum key size requirement of 1024 bits. Similarly, systems based on discrete logs involve the difficulty of calculating discrete logarithms in a finite field (Section 3.5.1). In these systems, key sizes are comparable to those in RSA. During the mid-1980s, cryptographers started studying the rich mathematical structures of elliptic curves and their applicability in primality testing and integer factorization (Lenstra 1987; Menezes 1993). In 1985, Neal Koblitz and Victor Miller proposed

## Table 3.1. Addition modulo 19 in $F_{19}$

| | 0 | 1 | 2 | 3 | 4 | 5 | 6 | 7 | 8 | 9 | 10 | 11 | 12 | 13 | 14 | 15 | 16 | 17 | 18 |
|---|---|---|---|---|---|---|---|---|---|---|---|---|---|---|---|---|---|---|---|
| 0 | 0 | 1 | 2 | 3 | 4 | 5 | 6 | 7 | 8 | 9 | 10 | 11 | 12 | 13 | 14 | 15 | 16 | 17 | 18 |
| 1 | 1 | 2 | 3 | 4 | 5 | 6 | 7 | 8 | 9 | 10 | 11 | 12 | 13 | 14 | 15 | 16 | 17 | 18 | 0 |
| 2 | 2 | 3 | 4 | 5 | 6 | 7 | 8 | 9 | 10 | 11 | 12 | 13 | 14 | 15 | 16 | 17 | 18 | 0 | 1 |
| 3 | 3 | 4 | 5 | 6 | 7 | 8 | 9 | 10 | 11 | 12 | 13 | 14 | 15 | 16 | 17 | 18 | 0 | 1 | 2 |
| 4 | 4 | 5 | 6 | 7 | 8 | 9 | 10 | 11 | 12 | 13 | 14 | 15 | 16 | 17 | 18 | 0 | 1 | 2 | 3 |
| 5 | 5 | 6 | 7 | 8 | 9 | 10 | 11 | 12 | 13 | 14 | 15 | 16 | 17 | 18 | 0 | 1 | 2 | 3 | 4 |
| 6 | 6 | 7 | 8 | 9 | 10 | 11 | 12 | 13 | 14 | 15 | 16 | 17 | 18 | 0 | 1 | 2 | 3 | 4 | 5 |
| 7 | 7 | 8 | 9 | 10 | 11 | 12 | 13 | 14 | 15 | 16 | 17 | 18 | 0 | 1 | 2 | 3 | 4 | 5 | 6 |
| 8 | 8 | 9 | 10 | 11 | 12 | 13 | 14 | 15 | 16 | 17 | 18 | 0 | 1 | 2 | 3 | 4 | 5 | 6 | 7 |
| 9 | 9 | 10 | 11 | 12 | 13 | 14 | 15 | 16 | 17 | 18 | 0 | 1 | 2 | 3 | 4 | 5 | 6 | 7 | 8 |
| 10 | 10 | 11 | 12 | 13 | 14 | 15 | 16 | 17 | 18 | 0 | 1 | 2 | 3 | 4 | 5 | 6 | 7 | 8 | 9 |
| 11 | 11 | 12 | 13 | 14 | 15 | 16 | 17 | 18 | 0 | 1 | 2 | 3 | 4 | 5 | 6 | 7 | 8 | 9 | 10 |
| 12 | 12 | 13 | 14 | 15 | 16 | 17 | 18 | 0 | 1 | 2 | 3 | 4 | 5 | 6 | 7 | 8 | 9 | 10 | 11 |
| 13 | 13 | 14 | 15 | 16 | 17 | 18 | 0 | 1 | 2 | 3 | 4 | 5 | 6 | 7 | 8 | 9 | 10 | 11 | 12 |
| 14 | 14 | 15 | 16 | 17 | 18 | 0 | 1 | 2 | 3 | 4 | 5 | 6 | 7 | 8 | 9 | 10 | 11 | 12 | 13 |
| 15 | 15 | 16 | 17 | 18 | 0 | 1 | 2 | 3 | 4 | 5 | 6 | 7 | 8 | 9 | 10 | 11 | 12 | 13 | 14 |
| 16 | 16 | 17 | 18 | 0 | 1 | 2 | 3 | 4 | 5 | 6 | 7 | 8 | 9 | 10 | 11 | 12 | 13 | 14 | 15 |
| 17 | 17 | 18 | 0 | 1 | 2 | 3 | 4 | 5 | 6 | 7 | 8 | 9 | 10 | 11 | 12 | 13 | 14 | 15 | 16 |
| 18 | 18 | 0 | 1 | 2 | 3 | 4 | 5 | 6 | 7 | 8 | 9 | 10 | 11 | 12 | 13 | 14 | 15 | 16 | 17 |

## Table 3.2. Multiplication modulo 19 in $F_{19}$

| | 0 | 1 | 2 | 3 | 4 | 5 | 6 | 7 | 8 | 9 | 10 | 11 | 12 | 13 | 14 | 15 | 16 | 17 | 18 |
|---|---|---|---|---|---|---|---|---|---|---|---|---|---|---|---|---|---|---|---|
| 0 | 0 | 0 | 0 | 0 | 0 | 0 | 0 | 0 | 0 | 0 | 0 | 0 | 0 | 0 | 0 | 0 | 0 | 0 | 0 |
| 1 | 0 | 1 | 2 | 3 | 4 | 5 | 6 | 7 | 8 | 9 | 10 | 11 | 12 | 13 | 14 | 15 | 16 | 17 | 18 |
| 2 | 0 | 2 | 4 | 6 | 8 | 10 | 12 | 14 | 16 | 18 | 1 | 3 | 5 | 7 | 9 | 11 | 13 | 15 | 17 |
| 3 | 0 | 3 | 6 | 9 | 12 | 15 | 18 | 2 | 5 | 8 | 11 | 14 | 17 | 1 | 4 | 7 | 10 | 13 | 16 |
| 4 | 0 | 4 | 8 | 12 | 16 | 1 | 5 | 9 | 13 | 17 | 2 | 6 | 10 | 14 | 18 | 3 | 7 | 11 | 15 |
| 5 | 0 | 5 | 10 | 15 | 1 | 6 | 11 | 16 | 2 | 7 | 12 | 17 | 3 | 8 | 13 | 18 | 4 | 9 | 14 |
| 6 | 0 | 6 | 12 | 18 | 5 | 11 | 17 | 4 | 10 | 16 | 3 | 9 | 15 | 2 | 8 | 14 | 1 | 7 | 13 |
| 7 | 0 | 7 | 14 | 2 | 9 | 16 | 4 | 11 | 18 | 6 | 13 | 1 | 8 | 15 | 3 | 10 | 17 | 5 | 12 |
| 8 | 0 | 8 | 16 | 5 | 13 | 2 | 10 | 18 | 7 | 15 | 4 | 12 | 1 | 9 | 17 | 6 | 14 | 3 | 11 |
| 9 | 0 | 9 | 18 | 8 | 17 | 7 | 16 | 6 | 15 | 5 | 14 | 4 | 13 | 3 | 12 | 2 | 11 | 1 | 10 |
| 10 | 0 | 10 | 1 | 11 | 2 | 12 | 3 | 13 | 4 | 14 | 5 | 15 | 6 | 16 | 7 | 17 | 8 | 18 | 9 |
| 11 | 0 | 11 | 3 | 14 | 6 | 17 | 9 | 1 | 12 | 4 | 15 | 7 | 18 | 10 | 2 | 13 | 5 | 16 | 8 |
| 12 | 0 | 12 | 5 | 17 | 10 | 3 | 15 | 8 | 1 | 13 | 6 | 18 | 11 | 4 | 16 | 9 | 2 | 14 | 7 |
| 13 | 0 | 13 | 7 | 1 | 14 | 8 | 2 | 15 | 9 | 3 | 16 | 10 | 4 | 17 | 11 | 5 | 18 | 12 | 6 |
| 14 | 0 | 14 | 9 | 4 | 18 | 13 | 8 | 3 | 17 | 12 | 7 | 2 | 16 | 11 | 6 | 1 | 15 | 10 | 5 |
| 15 | 0 | 15 | 11 | 7 | 3 | 18 | 14 | 10 | 6 | 2 | 17 | 13 | 9 | 5 | 1 | 16 | 12 | 8 | 4 |
| 16 | 0 | 16 | 13 | 10 | 7 | 4 | 1 | 17 | 14 | 11 | 8 | 5 | 2 | 18 | 15 | 12 | 9 | 6 | 3 |
| 17 | 0 | 17 | 15 | 13 | 11 | 9 | 7 | 5 | 3 | 1 | 18 | 16 | 14 | 12 | 10 | 8 | 6 | 4 | 2 |
| 18 | 0 | 18 | 17 | 16 | 15 | 14 | 13 | 12 | 11 | 10 | 9 | 8 | 7 | 6 | 5 | 4 | 3 | 2 | 1 |

**Table 3.3. Binary Polynomials for $F_{2^3}$**

|   | Binary polynomials for $F_{2^3}$ |
|---|---|
| 0 | 0 |
| 1 | 1 |
| 2 | x |
| 3 | x + 1 |
| 4 | $x^2$ |
| 5 | $x^2 + 1$ |
| 6 | $x^2 + x$ |
| 7 | $x^2 + x + 1$ |

a public-key cryptographic system using elliptic curves (Koblitz 1987). The algebraic structure of these elliptic curves forms the basis for elliptic curve cryptography. The fundamental security of this scheme relies on the difficulty of the discrete log problem in the elliptic curve setting (Section 3.7). Ever since this invention, there has been an enormous amount of interest in elliptic curve cryptography from diverse sectors of the research fraternity (Cohen and Frey 2006). In the early 1990s, they played a crucial role in the proof of Fermat's Last Theorem. Subsequently, elliptic curves were used in the definition of public-key crypto systems and opened a new paradigm in public key cryptography. ECC has been a proven technology and has been adopted by many standardizing bodies such as the National Institute of Standards and Technology (NIST), the International Organization for Standardization (ISO), the American National Standards Institute (ANSI), and the Standards for Efficient Cryptography Group (SECG). Furthermore, it has been incorporated in many commercial products such as e-mail systems, smart cards, and many applications in mobile technology.

### 3.5.1 Discrete log Problem

Let $(G, \times)$ be a multiplicative cyclic group with generator $g$ and order $n$. Let $x \in Z$ (private key) be an integer randomly selected from the interval $[1, n-1]$. Let $y$ be the public key such that

$$y = g^x \qquad (3.12)$$

The *discrete logarithm problem* in $G$ is defined as the problem of determining $x$ given $y, g$ and $n$.

## 3.5.2 Weierstrass Equations

The nonsingular Weierstrass equation is defined over a field $K$ as

$$y^2 + a_1 xy + a_3 y = x^3 + a_2 x^2 + a_4 x + a_6 \tag{3.13}$$

where $a_1, a_2, a_3, a_4$ and $a_6 \in K$. The set $K$ consists of the points $(x, y) \in K \times K$ along with *point to infinity* $(\mathcal{O})$ (We can imagine $\mathcal{O}$ in $K$ as seated infinitely far up the $y$ axis). It is best explained using projective space (Washington 2003). If $K$ is a finite field $F_q$, where $q$ is the order and having characteristic $p$, and #E(K) denotes the number of elements in $F_q$, then Hasse's theorem provides the tight bound for the cardinality of $E(K)$:

$$\left(\sqrt{q}-1\right)^2 \le \#\mathrm{E(K)} \le \left(\sqrt{q}+1\right)^2 \tag{3.14}$$

The discriminant $\Delta$ is

$$\Delta = -p_2^2 p_8 - 8 p_4^3 - 27 p_6^2 + 9 p_2 p_4 p_6 \tag{3.15}$$

where

$$p_2 = a_1^2 + 4a_2 \tag{3.16}$$

$$p_4 = 2a_4 + a_1 a_3 \tag{3.17}$$

$$p_6 = a_3^2 + 4a_6 \tag{3.18}$$

$$p_8 = a_1^2 a_6 + 4a_2 a_6 - a_1 a_3 a_4 + a_2 a_3^2 - a_4^2 \tag{3.19}$$

This generalized Weierstrass equation is more useful in binary fields. Hence,

$$\# E(K) = q + 1 - t \quad \text{where} \quad |t| \le 2\sqrt{q} \tag{3.20}$$

If $p$ is divisible by $t$, then $E$ is said to be supersingular; otherwise it is ordinary. However, if $|t| \le 2\sqrt{q}$ and $p$ is not divisible by $t$, then there exists an elliptic curve $E$ over $F_q$ with $E(F_q) = q + 1 - t$.

The Weierstrass equation for elliptic curve $E$ is defined as

$$y^2 = x^3 + Ax + B \tag{3.21}$$

where $A$, $B$, $x$, and $y$ are taken to be elements of a field. Hence, the discriminant is

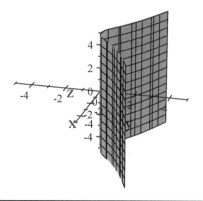

**Figure 3.1.** $E_1 : y^2 = x^3 + 2x$ Defined over $R$(3-D View)

$$\Delta = -(4A^3 + 27B^2) \tag{3.22}$$

We assume that $4A^3 + 27B^2 \neq 0$. Therefore the roots must be distinct.

As examples of elliptic curves, Figures 3.1 and 3.2 show $y^2 = x^3 + 2x$ over real numbers $(R)$.

Figures 3.3 and 3.4 show $y^2 = x^3 - 3x$ over real numbers.

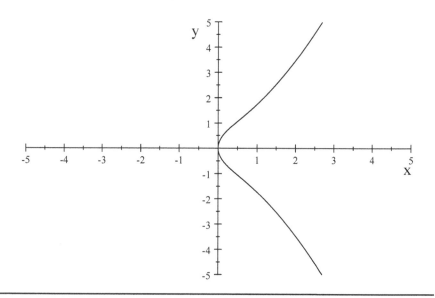

**Figure 3.2.** $E_1 : y^2 = x^3 + 2x$ Defined over $R$(2-D View)

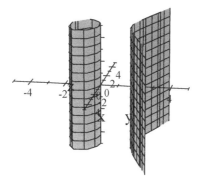

**Figure 3.3.**  $E_2 : y^2 = x^3 - 3x$  Defined over $R$(3-D View)

### 3.5.3 Prime Curves

Prime curves are defined over $Z_p$ (integer group from 0 to $p-1$), where $p$ is a large prime. Hence, limiting the variables and coefficients of the Weierstrass equation to a finite field $Z_p$, we have

$$y^2 \bmod p = \left(x^3 + Ax + B\right) \bmod p \tag{3.23}$$

For example, Figure 3.5 shows the equation $y^2 = x^3 + 2x$ over finite field $Z_p$, where $p = 23$.

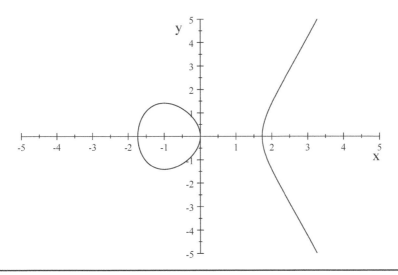

**Figure 3.4.**  $E_2 : y^2 = x^3 - 3x$  Defined over $R$(2-D View)

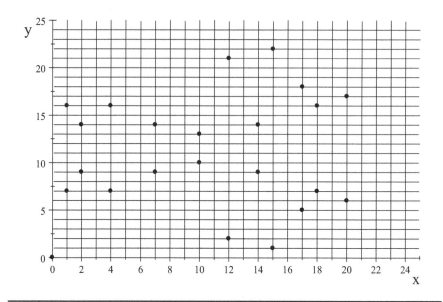

**Figure 3.5.**  $y^2 \bmod p = \left( x^3 + 2x \right) \bmod p$, where $p$ = 23

$$y^2 \bmod p = \left( x^3 + 2x \right) \bmod p \tag{3.24}$$

Points of elliptic curve  $y^2 \bmod p = \left( x^3 + 2x \right) \bmod p$, where $p$ = 23, are

$\{(0,0),(1,16),(1,7),(2,9),(2,14),(4,16),(4,7),(7,9),(7,14),(10,13),(10,10),$
$(12,2),(12,21),(14,9),(14,14),(15,1),(15,22),(17,18),(17,5),(18,16),(18,7),\mathcal{O}\}$

where $\mathcal{O}$ refers to *point to infinity*.

Basic operations include addition, subtraction, multiplication, division, and inversion.

### 3.5.4 Binary Curves

Binary curves are defined over the finite field $F(2^m)$, where the field consists of $2^m$ elements. In this case we use a cubic equation of the form

$$E_{2^m}(a,b): y^2 + xy = x^3 + ax^2 + b \tag{3.25}$$

where the coefficients and variables take on the values in $GF(2^m)$.

### 3.5.5 Group Law

If $E$ is an elliptic curve $y^2 = x^3 + Ax + B$ defined over the field $K$, then there exists a chord-and-tangent rule for performing an addition operation over the elements of the field $K$. Since the set of points $E(K)$ forms an abelian group where $\mathcal{O}$ serves as its identity, addition of two points in $E(K)$ will result in a third point in $E(K)$. Figure 3.6 shows a geometric representation of the addition rule. Let $P_1 = (x_1, y_1)$ and $P_2 = (x_2, y_2)$ be two distinct points on an elliptic curve $E$. Then the sum $P_3 = (x_3, y_3)$ can be obtained by drawing a line between $P_1$ and $P_2$, which will intersect the curve on $P_3'$. Reflection of $P_3'$ about the $x$ axis results in $P_3$.

In addition, Figure 3.7 shows the geometric representation of doubling a $P_1$. In this case, we draw a tangent line to the elliptic curve on $P$. This line intersects the curve at point $-R$. The result of $R$ is the reflection of $-R$ along the $x$ axis. As a special case, if the line is a tangent, it is assumed that it intersect at the point to infinity (Figure 3.8). In such cases, $2P = \mathcal{O}$.

Cryptographic applications use two families of elliptic curves: *prime curves* and *binary curves*.

#### 3.5.5.1 Group Law for $E / K : y^2 = x^3 + ax + b$, char$(K) \neq 2,3$

1.  Point addition. Let $P_1 = (x_1, y_1) \in E(K)$ and $P_2 = (x_2, y_2) \in E(K)$, where $P_1 \neq \pm P_2$. Then $P_1 + P_2 = (x_3, y_3)$, where

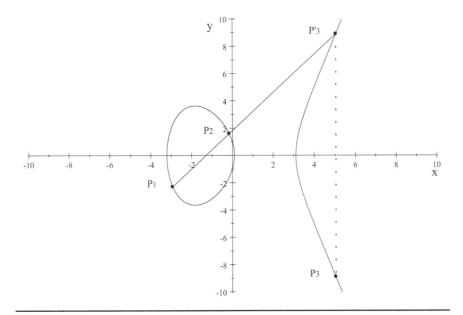

**Figure 3.6.** Geometric Representation of Addition Rule over an Elliptic Curve $E$

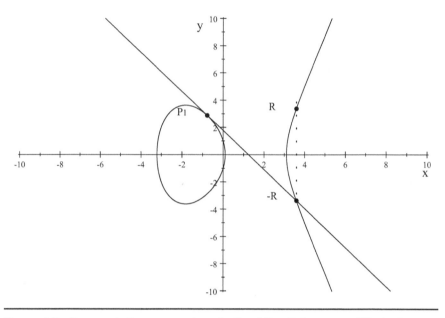

**Figure 3.7.** Geometric Representation of a Doubling Rule over an Elliptic Curve *E*

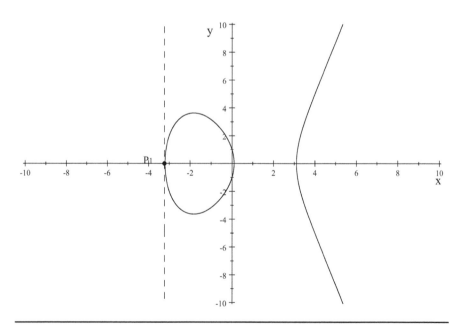

**Figure 3.8.** Geometric Representation When $2.P_1 = \mathcal{O}$ on *E*

$$x_3 = \left(\frac{y_2 - y_1}{x_2 - x_1}\right)^2 - x_1 - x_2 \tag{3.26}$$

$$y_3 = \left(\frac{y_2 - y_1}{x_2 - x_1}\right)(x_1 - x_3) - y_1 \tag{3.27}$$

2. Point doubling. Let $P_1 = (x_1, y_1) \in E(K)$, where $P_1 \neq -P_1$. Then $2P_1 = (x_3, y_3)$, where

$$x_3 = \left(\frac{3x_1^2 + a}{2y_1}\right)^2 - 2x_1 \tag{3.28}$$

and

$$y_3 = \left(\frac{3x_1^2 + a}{2y_1}\right)(x_1 - x_3) - y_1 \tag{3.29}$$

3. Identity. $P_1 + \infty = \infty + P_1 = P_1$ for all $P_1 \in E(K)$.
4. Negatives. If $P_1 = (x, y) \in E(K)$, then $(x, y) + (x, -y) = \infty$. The point $(x, -y)$ is denoted by $-P_1$ and is called the negative of $P_1$; note that $-P_1$ is indeed a point in $E(K)$. Also, $-\infty = \infty$.

Example of addition of two points on $E$ over $R$: Referring to Figure 3.4, where $P_1 = (-2.93, -2.3)$ and $P_2 = (-0.154, 1.6)$, then $P_3 = P_1 + P_2 = (5.05, -8.89)$.

Example of doubling of points on $E$ over $R$: Referring to Figure 3.5, where $P_1 = (-0.77, 2.88)$, then $2.P_1 = (3.6, 3.38)$.

Example of addition of two points on elliptic curve $y^2 = (x^3 + 2x)$ over $F_{23}$: Referring to Figure 3.3, let $P_1 = (4, 7)$ and $P_2 = (10, 10)$, then $P_3 = P_1 + P_2 = (15, 22)$.

Example of doubling of a point on elliptic curve $y^2 = (x^3 + 2x)$ over $F_{23}$: Referring to Figure 3.3, let $P_1 = (4, 7)$, then $2.P_1 = (1, 7)$.

## 3.5.5.2 Group Law for Nonsupersingular $E / F_{2^m}$ : $y^2 + xy = x^3 + ax^2 + b$

1. Point addition. Let $P_1 = (x_1, y_1) \in E(F_{2^m})$ and $P_2 = (x_2, y_2) \in E(F_{2^m})$, where $P_1 \neq \pm P_2$. Then $P_1 + P_2 = (x_3, y_3)$, where

$$x_3 = \lambda^2 + \lambda + x_1 + x_2 + a \tag{3.30}$$

and

$$y_3 = \lambda(x_1 + x_3) + x_3 + y_1 \tag{3.31}$$

with

$$\lambda = \frac{y_1 + y_2}{x_1 + x_2}$$

2.  Point doubling. Let $P_1 = (x_1, y_1) \in E(F_{2^m})$, where $P_1 \neq -P_1$. Then $2P_1 = (x_3, y_3)$, where

$$x_3 = \lambda^2 + \lambda + a = x_1^2 + \frac{b}{x_1^2} \tag{3.32}$$

and

$$y_3 = x_1^2 + \lambda x_3 + x_3 \tag{3.33}$$

3.  Identity. $P_1 + \infty = \infty + P_1 = P_1$ for all $P_1 \in E(F_{2^m})$
4.  Negatives. If $P_1 = (x, y) \in E(F_{2^m})$, then $(x, y) + (x, x + y) = \infty$. The point $(x, x + y)$ is denoted by $-P_1$ and is called the negative of $P_1$; note that $-P_1$ is indeed a point in $E(F_{2^m})$. Also, $-\infty = \infty$.

## 3.6  Why Elliptic Curve Cryptography?

One of the main reasons for using elliptic curves in cryptography is because of its reduction in key size, while providing the same level of security as RSA. For example, a 313-bit key size in elliptic curve systems will provide the same level of security as a 512-bit key size in RSA. Hence, it leads to faster computations as well as reduced storage space, processing power, and bandwidth.

## 3.7  Elliptic Curve Discrete Log Problem

Let $E$ be an elliptic curve defined over a finite field with $q$ elements $(F_q)$, such that $M \in F_q$ of order $n$. Let $N \in \langle P \rangle$ and

$$M = lN \tag{3.34}$$

where $l \in Z$ and $l \in [0, n-1]$. Hence, integer $l$ is called the discrete logarithm of $M$ to the base $N$, where $l = \log_N M$. (Note: The value of $l$ is very large in real cryptographic applications.)

In the area of solving hard problems, we normally gauge in terms of the size of the problem. For example, in the case of RSA, the hardness exists in factorizing the length of the modulus. In the case of elliptic curve crypto systems, the hardness exists in working with the number of points $q$ in the group.

## 3.8 Torsion Points

Every point on an elliptic curve $E$ has a finite order or an infinite order. Let $P \in E(\bar{K})$, where $\bar{K}$ is the algebraic closure of $K$, and let $n$ be a small integer; then there exists a finite subgroup such that

$$E_{tor}[n] = \{P \in E(K) \mid nP = \mathcal{O}\} \subseteq E[n] \qquad (3.35)$$

Hence, $n$ is the order of $P$. On the contrary, if no such $n$ exists for a $P \in E(\bar{K})$ (we will never get $\mathcal{O}$ by adding $P$ to itself), then the order of the point is infinite.

**Example 3.1.** Let us consider the elliptic curve $E / K : y^2 = x^3 + 3x$, where $K = F_{11}$ (Figure 3.9).

The set of rational points on $E$ is $\{[0,0],[1,9],[1,2],[2,5],[2,6],[3,5],[3,6],[6,5],[6,6],$ $[7,1],[7,10],\mathcal{O}\}$. Therefore, $\#E(K) = 12$. Referring to Eq. (3.20), the value of $t = 0$. Hence the elliptic curve $E / K : y^2 = x^3 + 3x$ defined over $F_{11}$ is supersingular. Table 3.4 shows the order of the torsion points in $E / K : y^2 = x^3 + 3x$ where $K = F_{11}$.

Let $E : y^2 = x^3 + Ax + B$ be an elliptic curve and let $(r_1, r_2, r_3) \in \bar{K}$ be the roots of the cubic equation $E$. Then

$$y^2 = (x - r_1)(x - r_2)(x - r_3) \qquad (3.36)$$

If a point $P$ reaches infinity at $2P$, then a line through the $P$ is a tangent (Figure 3.5). Hence, we get the subgroup with second-order torsion points.

$$E_{tor}[2] = \{(r_1, 0), (r_2, 0), (r_3, 0), \mathcal{O}\} \qquad (3.37)$$

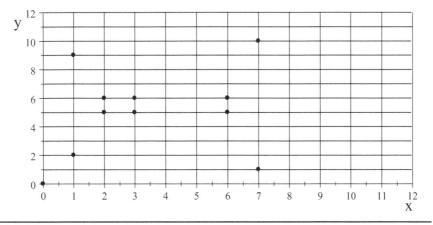

**Figure 3.9.** $E / K : y^2 = x^3 + 3x$, where $K = F_{11}$

**Table 3.4. Finite Subgroup $E_{tor}[n]$ for $E/K : y^2 = x^3 + 3x$,**
**Where $K = F_{11}$**

| Point P | Order n | Point P | Order n |
|---------|---------|---------|---------|
| [0,0]   | 2       | [3,6]   | 3       |
| [1,9]   | 6       | [6,5]   | 4       |
| [1,2]   | 6       | [6,6]   | 4       |
| [2,5]   | 12      | [7,1]   | 12      |
| [2,6]   | 12      | [7,10]  | 12      |
| [3,5]   | 3       | $\mathcal{O}$ | 1 |

Let us review Mazur's theorem, by which we limit the number of groups to a rational torsion subgroup.

*Mazur's theorem:* Let $E$ be a nonsingular cubic elliptic curve, and let $E(\mathbb{Q})$ contain a point $P$ of finite order $m$. Then the torsion group is isomorphic to $Z/nZ$ for $1 \le n \le 10$ or $n = 12$ or $Z/2Z \times Z/2nZ$ for $1 \le n \le 4$.

## 3.9 Pairing-Based Cryptography

In recent years we have seen enormous usage of pairings in cryptography, which have been particularly useful in implementing identity-based signature and encryption schemes (Boneh and Boyen 2004; Boneh and Franklin 2001; Boneh and Shacham 2004; Boneh et al. 2004; Joux 2000; Sakai et al. 2000). In this section, we review Weil and Tate pairings.

### 3.9.1 Divisor Theory

A *divisor* is the formal sum of points on the curve $E : y^2 = x^3 + Ax + B$ defined over a finite field $F_p$.

$$A = \sum_i a_i[P_i] \quad \text{where} \quad a_i \in Z \text{ and } P_i \in E \quad (3.38)$$

The *degree* of the divisor is defined as

$$\deg\left(\sum_i a_i[P_i]\right) = \sum_i a_i \in Z \quad (3.39)$$

$$\text{sum}\left[\sum_i a_i [P_i]\right] = \sum_i a_i P_i \qquad (3.40)$$

A divisor $A = \sum_i a_i [P_i]$ is principal if and only if $\deg\left(A = \sum_i a_i [P_i]\right) = 0$ and $\text{sum}\left(A = \sum_i a_i [P_i]\right) = \infty$. A group of divisors of $E$, denoted as $\text{Div}(E)$, is an abelian group formed by the points on $E$. In addition, divisors of degree 0 $[\text{Div}^0(E)]$ form a subgroup of $\text{Div}(E)$, which is of particular interest to us.

### 3.9.2 Rational Function

A function on the curve $E$ can be viewed as a rational function $f(x,y) \in \bar{K}(x,y)$ if it is defined for at least one point in $E$. Let $P(x,y) \in E$; then we define $f(P) = f(x,y)$.

**Example 3.2.** Let $E$ be a curve $y^2 = x^3 + x$ and let $f(x,y) = x/y$. Note that the function $f(x,y)$ is not defined for $P = (0,0)$. However, on the curve $E$, the function $f(x,y)$ is

$$f(x,y) = \frac{x}{y} = \frac{y}{x^2 + 1} \qquad (3.41)$$

and equates to 0 $P$. Furthermore, the $f(x,y)$ can be transformed to

$$f(x,y) = \frac{y}{x} = \frac{y}{x^2 + 1} \qquad (3.42)$$

and becomes $\infty$ at $P$.

### 3.9.2.1 Zeros and Poles

A rational function on the curve is said to have a *zero* at point $P$ if the function takes the value 0 and point $P$, $f(P) = 0$. And a function on the curve is said to have a *pole* at point $P$ if the function takes the value $\infty$ at point $P$.

### 3.9.2.2 Order of Poles and Zeros

It can be shown that there is a function $u_p$, called the Uniformizer at $P$, with $u(P) = 0$, and such that every rational function $f(x,y)$ can be written in the form

$$f = \left(u_p\right)^r g \qquad (3.43)$$

with $r \in Z$ and $g(P) \neq 0, \infty$.

### 3.9.2.3 Divisor of a Function

Let $f$ be a function on a curve; then the divisor of the function is given by

$$\text{div}(f) = \sum_{P_i} \text{ord}_P(f).(P) \tag{3.44}$$

where $\text{ord}_p(f) \in Z$ and $(f)$ is a point on the curve $E$.

**Example 3.3.** Let us consider the same elliptic curve equation $E$ as in Example 3.2. Let $f(x, y)$ be a rational function such as

$$f(x, y) = x = y^2 \left( \frac{1}{x^2 + 1} \right) \tag{3.45}$$

Where $u_P = y = 0$ at point $P = (0, 0)$ and $g(P) = (1/x^2 + 1) \neq 0$, $\infty$ at point $P$. Hence, in this example, $\text{ord}_P(x) = 2$.

**Lemma 3.1.** Let $f$ be a rational function on an elliptic curve $E$ such that $f \neq 0$. Then

- $f$ has only a finite number of zeros and poles.
- $\deg[\text{div}(f)] = 0$.
- $f$ is a constant if the function does not have zeros and poles.

Let there be three points $(P_1, P_2, P_3)$ on $E$ and let a line $f_1(x, y) = ax + by + c = 0$ pass through these three points. Then this function $f_1$ has three zeros at $P_1, P_2, P_3$. And if $b \neq 0$, then $f_1$ has three poles at $\infty$. Hence, the divisor of the line function is

$$\text{div}(ax + by + c) = [P_1] + [P_2] + [P_3] - 3[\infty] \tag{3.46}$$

Let $-P_3$ be a reflection of $P_3(x_3, y_3)$. Then the line passing through $P_3$ and $-P_3$ is given by $f_2(x, y) = x - x_3 = 0$.
The divisor of $f_2$ is given by

$$\text{div}(x - x_3) = [P_3] + [-P_3] - 2[\infty] \tag{3.47}$$

Therefore,

$$\text{div}\left( \frac{ax + by + c}{x - x_3} \right) = \text{div}(ax + by + c) - \text{div}(x - x_3) \tag{3.48}$$

$$\text{div}\left( \frac{ax + by + c}{x - x_3} \right) = [P_1] + [P_2] - [-P_3] - [\infty] \tag{3.49}$$

Since $-P_3 = P_1 + P_2$ on $E$, the above equation can be rewritten as

$$[P_1]+[P_2]=[P_1 + P_2]+[\infty]+\underbrace{\mathrm{div}\left|\frac{ax+by+c}{x-x_0}\right|}_{g} \tag{3.50}$$

**Lemma 3.2.** Let $D$ be a divisor on $E$ with $\deg(D)=0$. Then there exists a function $f$ on $E$ such that

$$\mathrm{div}(f)= D \tag{3.51}$$

if and only if

$$\mathrm{sum}(D)= \infty \tag{3.52}$$

## 3.9.3 Weil Pairing

Weil pairing has existed since the introduction of Weierstrass's elliptic curve functions and was introduced by him as a sigma function on the elliptic curves. In 1940, André Weil gave a more abstract definition of this mapping in the Riemann hypothesis for arbitrary genus curves over finite fields. Weil pairing is best explained mathematically.

Let $E$ be an elliptic curve defined over a field $K$ and let $n$ be an integer such that it is not divisible by the characteristic of $K$. Let $z \in \bar{K}$ such that $z^n = 1$. Then we define a group of $n$th roots of unity in $\bar{K}$ such that

$$\mu_n = \{z \in \bar{K} \,|\, z^n = 1\} \tag{3.53}$$

The equation $z^n = 1$ does not have multiple roots, because $n$ is not divisible by the characteristic of $K$. Hence, the $n$ roots exist in $\bar{K}$ and $\mu_n$ is a cyclic group with order $n$. Every cyclic group has a generator, and let $\zeta$ be the generator for $\mu_n$. Here, $\zeta$ is called a primitive $n$th root of unity. Hence, *Weil pairing* is defined as a paring

$$e_n : E[n] \times E[n] \rightarrow \mu_n \tag{3.54}$$

Let $L \in E[n]$. Then, by Lemma 3.2, there exists a function such that

$$\mathrm{div}(f) = n[L] - n[\infty] \tag{3.55}$$

Let us choose $L' \in E[n^2]$ such that $nL' = L$. Referring to Eq. (3.50), $\mathrm{div}(g)$ is given as

$$\mathrm{div}(g) = \sum_{M \in E[n]} ([L' + M] - [M]) \tag{3.56}$$

where the sum of the points in div($g$) equals $\infty$. Hence,

$$\text{div}(g) = \sum_{nL'=L} [L'] + \sum_{nM=\infty} [M] \qquad (3.57)$$

Let $f \circ n$ denote the function such that it starts with a point, $n$ is multiplied with the point, and then the function is applied. Let $P = L' + M$ with $M \in E[n]$ those points $P$ with $nP = L$. Hence, it follows from Eq. (3.55) that

$$\text{div}(f \circ n) = n\left(\sum_M [L' + M]\right) - n\left(\sum_M [M]\right) = \text{div}(g^n) \qquad (3.58)$$

Therefore,

$$f \circ n = g^n \qquad (3.59)$$

Let $P \in E(\bar{K})$ and $Q \in E[n]$. Then

$$g(P+Q)^n = f(n(P+Q)) = f(nP) = g(P)^n \qquad (3.60)$$

Hence, we define Weil pairing as

$$e_n(Q,L) = \frac{g(P+Q)}{g(P)} \in \mu_n \qquad (3.61)$$

### 3.9.4 Bilinear Property

Let $G_1$ and $G_2$ be two groups of order $q$, where $q$ is a large prime. Then the bilinear map $e : G_1 \times G_2 \rightarrow G_2$ satisfies the following properties.

1.  Bilinear: The map $e : G_1 \times G_2 \rightarrow G_2$ is bilinear if

    $$e(aP,bQ) = e(P,Q)^{ab} \ \forall \, a,b \in Z \qquad \text{and} \qquad P,Q \in G_1 \qquad (3.62)$$

2.  Nondegenerate: All pairs in $G_1 \times G_1$ do not map to the identity of $G_2$. Let $P$ be the generator of $G_1$. Then $e(P,P)$ is a generator of $G_2$, because $G_1, G_2$ are prime order groups.
3.  Efficient algorithms exist to compute $e(P,Q)$, where $P,Q \in G_1$.

### 3.9.5 Tate Pairing

Let $E$ be an elliptic curve defined over the finite field $F_q$, and let $n$ be an integer such that $n$ is divisible by $q-1$. Let the elements of $F_q$ whose order is divisible by $n$ be denoted as $E\left(F_q\right)[n]$ and let $\mu_n = \{z \in \bar{K} \mid z^n = 1\}$. Let us assume that $E\left(F_q\right)$ contains elements of order $n$. Then there exists a *modified Tate-Lichtenbaum pairing,*

$$\tau_n : E\left(F_q\right)[n] \times E\left(F_q\right)[n] / nE\left(F_q\right)[n] \rightarrow \mu_n \tag{3.63}$$

## 3.10  Summary

We have introduced some basic concepts of modern algebra and looked into some basic definitions in field theory. We discussed the importance of elliptic curve cryptography and then looked into the elliptic curve discrete log problem. We then looked into pairing-based cryptography and, in particular, Weil and Tate pairing.

## 3.11  References

Boneh, D., and X. Boyen. Efficient selective-ID secure identity-based encryption without random oracles. *Lecture Notes in Computer Science* (Springer-Verlag) 3027 (2004): 223–238.

Boneh, D., and M. Franklin. Identity-based encryption from theWeil pairing. *Lecture Notes in Computer Science* (Springer-Verlag) 2139 (2001): 213–229.

Boneh, D., and H. Shacham. Group signatures with verifier-local revocation. *ACM Conference on Computer and Communications Security.* ACM, 2004, 168–177.

Boneh, D., B. Lynn, and H. Shacham. Short signatures from the Weil pairing. *Journal of Cryptology* (2004): 297–319.

Cohen, H., and G. Frey. *Elliptic and Hyperelliptic Curve Cryptography.* Edited by K. H. Rosen. Chapman & Hall/CRC Press, 2006.

Diffie, W., and M. E. Hellman. New directions in cryptography. *IEEE Transactions on Information Theory* (IEEE) IT-22 (1976): 644–654.

ElGamal, T. A public-key cryptosystem and a signature scheme based on discrete logarithms. *IEEE Transactions on Information Theory* (IEEE) 31, no. 4 (1984): 469–472.

Joux, A. A one round protocol for tripartite Diffie–Hellman. *Lecture Notes in Computer Science* (Springer-Verlag) 1838 (2000): 385–394.

Koblitz, N. Elliptic curve cryptosystems. *Mathematics of Computation* 48, no. 177 (1987): 203–209.

Lenstra, H. W. Factoring integers with elliptic curves. *Annals of Mathematics* 126, no. 3 (1987): 649–673.

Menezes, A. *Elliptic Curve Public Key Cryptosystems.* Kluwer Academic, 1993.

Rivest, R., A. Shamir, and L. Adleman. A method for obtaining digital signatures and public-key cryptosystems. *Communications of the ACM* (ACM) 21, no. 2 (1978): 120–126.

Sakai, R., K. Ohgishi, and M. Kasahara. Cryptosystems based on pairing. *The 2000 Symposium on Cryptography and Information Security.*

Simmons, G. J. *Contemporary Cryptology: The Science of Information Integrity.* Wiley-IEEE Press, 1999.

Washington, L. C. *Elliptic Curves: Number Theory and Cryptography.* Chapman & Hall/CRC Press, 2003.

# Chapter 4

## Identity-Based Cryptography

### In This Chapter

- Introduction to public key cryptography
- Identity-based cryptography
- Review of identity-based authentication and encryption schemes
- Key escrow problem and threshold signature schemes

## 4.1 Introduction

Throughout the history of human civilization, the need for secure communication has been one of the top priorities. The word *cryptography* comes from two Greek words (*kryptos* = *hidden*) and (*graphein* = *writing*). In fact, the need to send concealed messages dates back to the Stone Age, when human beings started living in groups. As soon as these groups evolved, the level of trust among them deteriorated. Hence, the need to send messages among the allied groups/tribes in a hidden manner increased. As early as 1900 B.C., the Greeks wrapped a tape on a stick and wrote a message on the tape in a sequential manner. Upon unwinding the tape from the stick and reading the message, the text would be meaningless. To decipher the message, the receiver needed a stick with the same diameter as that of the sender. However, breaking this code was simply a matter of trying sticks with different diameters.

Julius Caesar invented one of the earliest cryptographic systems to send encrypted messages to his generals. It is popularly known as the Caesar shift cipher, wherein each letter in the message is shifted by a specific number of letters that has been agreed on by the sender and the receiver. This cipher is categorized under monoalphabetic ciphers and can be easily broken by listing the alphabet repeatedly, aligning each letter sequentially with succeeding letters, until the message is decrypted.

Most early cryptographic schemes can be categorized as symmetric key cryptographic systems. The Middle Ages did not see any major advancement in the art and science of cryptography. In 1467, Leon Battista Alberti invented the polyalphabetic cipher. This is simply a block cipher in which the plain text is mapped to one of several possibilities in the cipher text and vice versa. However, polyalphabetic ciphers went through a variety of changes, and the modern version is attributed to Vigenere. For example, let the keyword be "MEC" and the plain text be "We need more supplies fast." Using the Vigenere square table (Table 4.1), the Vigenere cipher is

| M | E | C | M | E | C | M | E | C | M | E | C | M | E | C | M | E | C | M | E | C | M |
|---|---|---|---|---|---|---|---|---|---|---|---|---|---|---|---|---|---|---|---|---|---|
| w | e | n | e | e | d | m | o | r | e | s | u | p | p | l | i | e | s | f | a | s | t |
| I | I | P | Q | I | F | Y | S | T | Q | W | W | B | T | N | U | I | U | R | E | U | F |

## Table 4.1. Vigenere Cipher Square Table

|    | a | b | c | d | e | f | g | h | i | j | k | l | m | n | o | p | q | r | s | t | u | v | w | x | y | z |
|----|---|---|---|---|---|---|---|---|---|---|---|---|---|---|---|---|---|---|---|---|---|---|---|---|---|---|
| 1  | B | C | D | E | F | G | H | I | I | J | K | L | M | N | O | P | Q | R | S | T | U | V | W | X | Y | Z | A |
| 2  | C | D | E | F | G | H | I | J | K | L | M | N | O | P | Q | R | S | T | U | V | W | X | Y | Z | A | B |
| 3  | D | E | F | G | H | I | J | K | L | M | N | O | P | Q | R | S | T | U | V | W | X | Y | Z | A | B | C |
| 4  | E | F | G | H | I | J | K | L | M | N | O | P | Q | R | S | T | U | V | W | X | Y | Z | A | B | C | D |
| 5  | F | G | H | I | J | K | L | M | N | O | P | Q | R | S | T | U | V | W | X | Y | Z | A | B | C | D | E |
| 6  | G | H | I | J | K | L | M | N | O | P | Q | R | S | T | U | V | W | X | Y | Z | A | B | C | D | E | F |
| 7  | H | I | J | K | L | M | N | O | P | Q | R | S | T | U | V | W | X | Y | Z | A | B | C | D | E | F | G |
| 8  | I | J | K | L | M | N | O | P | Q | R | S | T | U | V | W | X | Y | Z | A | B | C | D | E | F | G | H |
| 9  | J | K | L | M | N | O | P | Q | R | S | T | U | V | W | X | Y | Z | A | B | C | D | E | F | G | H | I |
| 10 | K | L | M | N | O | P | Q | R | S | T | U | V | W | X | Y | Z | A | B | C | D | E | F | G | H | I | J |
| 11 | L | M | N | O | P | Q | R | S | T | U | V | W | X | Y | Z | A | B | C | D | E | F | G | H | I | J | K |
| 12 | M | N | O | P | Q | R | S | T | U | V | W | X | Y | Z | A | B | C | D | E | F | G | H | I | J | K | L |
| 13 | N | O | P | Q | R | S | T | U | V | W | X | Y | Z | A | B | C | D | E | F | G | H | I | J | K | L | M |
| 14 | O | P | Q | R | S | T | U | V | W | X | Y | Z | A | B | C | D | E | F | G | H | I | J | K | L | M | N |
| 15 | P | Q | R | S | T | U | V | W | X | Y | Z | A | B | C | D | E | F | G | H | I | J | K | L | M | N | O |
| 16 | Q | R | S | T | U | V | W | X | Y | Z | A | B | C | D | E | F | G | H | I | J | K | L | M | N | O | P |
| 17 | R | S | T | U | V | W | X | Y | Z | A | B | C | D | E | F | G | H | I | J | K | L | M | N | O | P | Q |
| 18 | S | T | U | V | W | X | Y | Z | A | B | C | D | E | F | G | H | I | J | K | L | M | N | O | P | Q | R |
| 19 | T | U | V | W | X | Y | Z | A | B | C | D | E | F | G | H | I | J | K | L | M | N | O | P | Q | R | S |
| 20 | U | V | W | X | Y | Z | A | B | C | D | E | F | G | H | I | J | K | L | M | N | O | P | Q | R | S | T |
| 21 | V | W | X | Y | Z | A | B | C | D | E | F | G | H | I | J | K | L | M | N | O | P | Q | R | S | T | U |
| 22 | W | X | Y | Z | A | B | C | D | E | F | G | H | I | J | K | L | M | N | O | P | Q | R | S | T | U | V |
| 23 | X | Y | Z | A | B | C | D | E | F | G | H | I | J | K | L | M | N | O | P | Q | R | S | T | U | V | W |
| 24 | Y | Z | A | B | C | D | E | F | G | H | I | J | K | L | M | N | O | P | Q | R | S | T | U | V | W | X |
| 25 | Z | A | B | C | D | E | F | G | H | I | J | K | L | M | N | O | P | Q | R | S | T | U | V | W | X | Y |
| 26 | A | B | C | D | E | F | G | H | I | J | K | L | M | N | O | P | Q | R | S | T | U | V | W | X | Y | Z |

Vigenere also introduced the concept of the one-time pad into cryptography.

The proliferation of computer communication in the 1960s brought with it an increased demand to secure digital information. In addition, the involvement of the private sector in providing digital security led to the work on symmetric ciphers by Feistal at IBM that culminated in the most widely used encryption standard, known as Data Encryption Standard (DES). In 1977, the National Institute of Standards and Technology (NIST) adopted DES and published it as the Federal Information Processing Standard 46 (FIPS PUB 46). Since then, many financial institutions accepted DES until the emergence of the Advanced Encryption Standards (AES).

In 1976, Diffie and Hellman introduced a revolutionary concept in cryptography, which led to the emergence of public key cryptography. In the next two sections of this chapter, we review symmetric and asymmetric key cryptography.

### 4.1.1 Symmetric Key Cryptographic Systems

In symmetric key cryptography, the same key is used for encryption and decryption of messages. Figure 4.1 depicts the symmetric key cryptography system.

In this example, the sender (Alice) uses a key (k) and encrypts the message (M) to generate the cipher text (C). The cipher text is then sent over a public channel to the receiver (Bob). The receiver (Bob) decrypts the cipher text (C) by using the same key (k) and retrieves the message (M). However, this method has critical drawbacks, such as the following:

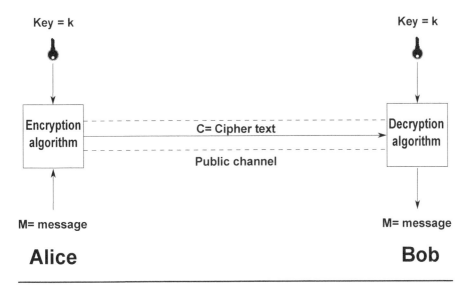

**Figure 4.1.** Symmetric Key Cryptographic Model

- The key must remain secret at both the sender's and the receiver's end, in a two-party communication.
- Complexity in managing keys arises with an increase in scalability.
- It requires an unconditionally trusted third-party key generator and distributer that is reachable by every node in the network.
- It is recommended that nodes refresh their keys periodically to abide with sound cryptographic practices. However, frequent queries to the key generator can be overwhelming and may lead to an unintentional denial of service.

With the advent of radio communication and with its high acceptance rate, the need for secure communications was evident. A drastic shift in the paradigm of cryptography took place with the introduction of asymmetric key cryptography.

### 4.1.2 Asymmetric Key Cryptographic Systems

This new class of cryptographic system introduced the concept of using two separate keys for encryption and decryption. That is, each party owns two keys, the public key and a corresponding private key. Figure 4.2 depicts an asymmetric key cryptographic system used for encrypting messages.

In this system, Alice uses Bob's public key $[B(pu_k)]$ to send encrypted messages and, correspondingly, Bob uses his private key $[B(pr_k)]$ to decrypt the message. This ground-

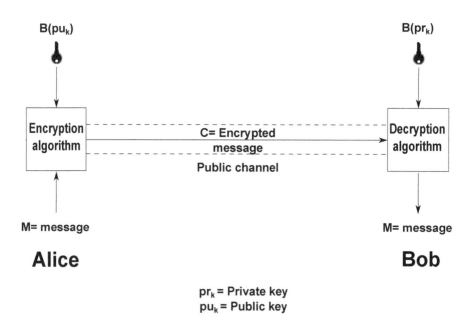

**Figure 4.2.** Encryption in Asymmetric Key Cryptography

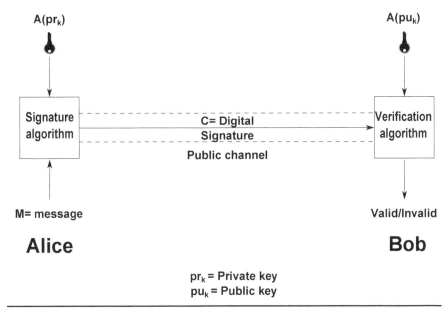

**Figure 4.3.** Digital Signatures in Asymmetric Key Cryptography

breaking approach of public key encryption was first introduced by Diffie and Hellman in their seminal paper titled "New Directions in Cryptography" (Diffie and Hellman 1976). However, they were unable to come up with a concrete mathematical proof of their proposed scheme. In 1978, three researchers (Rivest, Shamir, and Adleman) came up with a convincing algorithm that conceptualized public key encryption and called it the RSA algorithm (Rivest et al. 1978). Furthermore, the notion of a digital signature was introduced with the establishment of public key cryptographic systems. In this scheme, each end user uses his private key to encrypt a message, resulting in a digital signature. Likewise, the signature is verified using the end user's public key (Figure 4.3). Therefore, Alice uses her private key $[A(pr_k)]$ to encrypt a message that results in a digital signature. Any recipient of digital signature can verify the signature by using Alice's public key $[A(pu_k)]$. Hence, each end user is able to provide authenticity using the concept of digital signatures.

In 1984, ElGamal proposed a public key encryption scheme that was closely related to the Diffie-Hellman key agreement protocol (Elgamal 1985). In this scheme, the plain text message is mapped to a single group element while the cipher text is mapped to two group elements. Hence, the cipher text is longer than the plain-text message. At that time, the notion of having a larger cipher text compared to plain text was a difficult concept for the cryptographic community to digest. Also, in either of the schemes (RSA or Elgamal's signature scheme), two important questions arise:

- How and from what source does Bob retrieve Alice's public key?
- If Bob receives Alice's public key from Alice, how can Bob verify the binding between Alice's unique identity and the public key?

Hence, it was evident that public key crypto-systems introduced a new paradigm in cryptography, but also increased the complexity of securing end-to-end communications. Public Key Infrastructure (PKI) aims to solve the above questions.

## 4.2 Public Key Infrastructure

Public Key Infrastructure (PKI) was designed to provide the necessary foundation for implementing security services. One of the main goals of PKI is to provide an arrangement to bind the public keys of end users with their respective identities. In PKI, this binding is achieved through a digital certificate (For example, X.509 version 3 certificates). A *digital certificate* is a signed certificate that contains the public key along with the name of the subject. The name of the subject can be an email address, a directory name, or a Domain Name Service (DNS) name. These certificates are usually signed by certificate authorities (CAs). A *certificate authority* is supposedly a trusted third party who is responsible for issuing certificates to trusted users or intermediate certificate authorities. A typical PKI consists of CAs, Registration Authorities (RAs), certificate repository, and certificate management (discussed in the next section).

### 4.2.1 Single-Certificate Authority Model

In this model, every end user's equipment is preconfigured with the Certificate Authority (CA)'s public key certificate (Figure 4.4). The organizations that run CA (VeriSign or Thwate) are responsible for distributing public key certificates. However, there is no single organization that is trusted universally by every commercial, governmental, educational institution, and so on. From a scalability point of view, it is tedious, expensive, and unsecure to obtain certificates from unrelated and distant organizations.

In addition, adhering to the norm of best security practice; it is important to update the public key certificate of the CA periodically. However, with the single-certificate model, it is next to impossible to refresh public key certificates of all the end users at the

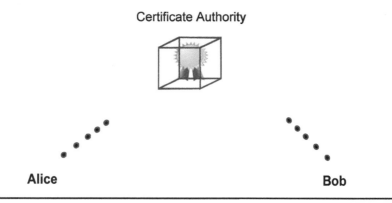

**Certificate Authority**

Alice                                                                 Bob

**Figure 4.4.** Single-Certificate Authority Model

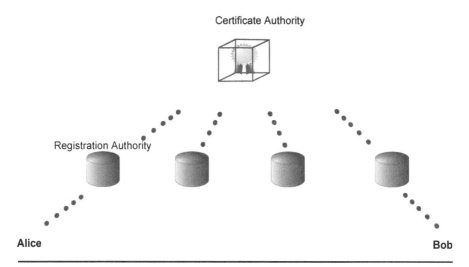

**Figure 4.5.** CA + RA Model

same time. Furthermore, the domination of a single organization responsible for distributing certificates might lead to a monopoly by one organization in this technology, resulting in excessive fees for distributing certificates. Hence, it is recommended that we have a hierarchical CA architecture instead of a single CA model (Figure 4.5). This recommendation could lead to multiple root CAs instead of a single CA.

A subset of the single CA model might include multiple Registration Authorities spread across different domains. Each Registration Authority (RA) is responsible for verifying the mapping between an end user's name and his public key. Upon authenticating each end user, the intermediate RAs send a signed message to the CA with the request to grant the public key certificate.

Although this model reduces the burden of having all work done by the CA, it is still a single CA model and inherits most of the drawbacks associated with the single CA model.

## 4.2.2 A Hierarchy of Certificate Authorities

In this model (Figure 4.6), each domain has a CA that is responsible for authenticating the mapping between an end user's identity and the public key. Furthermore, each domain can be subdivided into multiple CAs and RAs. A separate entity may take the role of an RA, or it can be combined with the CA.

Hence, this model facilitates a conducive environment for healthy competition among the organizations owning the root CAs. However, this model increases the security risk if any one of the root CA's private keys is compromised, leading to a complete failure of the system. In practice, a simple social engineering attack could trick less tech-savvy people to accept certificates and add them to their trusted list of CAs. However, in the world of wireless security networks, sensors can be preprogrammed not to accept any new certificates from a non-trusted source.

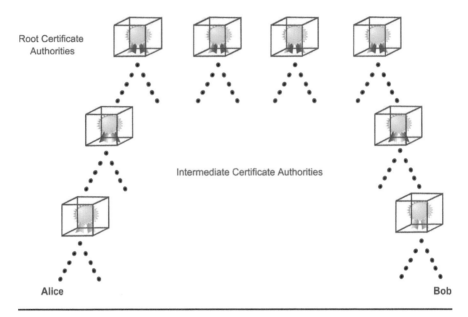

**Figure 4.6.** CA Hierarchy Model

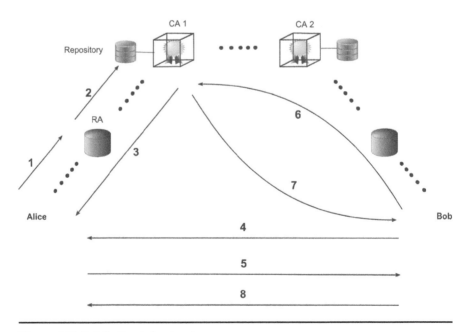

**Figure 4.7.** Public Key Infrastructure

Figure 4.7 portrays the basic PKI model that is used in many enterprises.

1. In this example, Alice must first enroll or register with the RA in order to participate in the PKI. Alice generates a private/public key pair and sends the public key along with the related identity information to the nearest RA.
2. The RA validates Alice's information and forwards the related information to its CA (CA1). In an enterprise environment, this process is tightly controlled and the entity that plays the role of an RA is assumed to be trusted.
3. The main objective of CA is to bind the identity information along with the credentials supplied by RA with Alice's public key. The binding is established when the CA digitally signs the public key certificate that is sent back to Alice.
4. If Bob is interested in communicating securely with Alice, he will need Alice's public key certificate. Hence, he sends a request to Alice to procure the public key certificate.
5. Alice responds to the request by sending her digital certificate.
6. Upon receiving the digital certificate, Bob needs to validate the certificate and sends this certificate to CA1.
7. The concerned CA (CA1) checks with the repository that stores the Certificate Revocation List (CRL) to see if the certificate has expired. In addition, Bob verifies the CA's digital signature by using his CA's public key embedded in the other root CA (CA1's certificate). If CA1's certificate is not already stored in Bob's cache, he needs to send a request to CA1 in setup 6.
8. Upon successfully verifying Alice's certificate, Bob can send an encrypted message to Alice.

## 4.2.3 Pros and Cons of PKI

Although PKI has existed for more than two decades, and has revolutionized security for online business models (e-commerce), it is by no means a panacea that facilitates seamless end-to-end security and allows the business world to accept it with utter confidence. Even though PKI provides a robust user authentication mechanism in the enterprise domain, major problems have been encountered in leveraging digital certificates in encryption. In the enterprise world, the preenrollment problem persists in PKI, where senders need to ensure that receivers have their certificates (usually stored in a repository) before sending an encrypted message to their receivers. First, the introduction of third parties maintaining public key certificates introduces problems such as trust. A receiver who is in the process of verifying the sender's digital signature will trust the root CA of that sender's domain. Second, the possibility of information leakage exists, and third, whether the binding between the certificate and binding was legitimate at the time of issuance may also be a concern.

In the context of WSN, the direct application of PKI presented in Figure 4.7 is not practical, as it would introduce extra communication overhead in a resource-constrained system. The lack of a trusted infrastructure in an ad-hoc, decentralized environment makes it difficult to apply the concept of certificate authority directly. Second, it is not possible for any of the resource-constrained nodes to play the role of a CA, as the CA should be unconditionally trusted by all the nodes in the network and should have a tamper-resistant material to prevent any kind of node-compromising attack. Although the sink node could act as a CA and as an RA in a WSN environment, it would be very energy inefficient in a hop-by-hop routing model for each node to retrieve certificates from the sink. Furthermore, it would lead to resource exhaustion and a significant reduction in network lifetime. While preloading of digital certificates in nodes before deployment has been proposed, it would lead to unnecessarily large storage overhead and inefficient resource utilization. In addition, certificate management in WSN would be very complex, as it would need to go through a complex path construction process that would generate yet more communication overhead in the network.

## 4.3 Identity-Based Cryptography

The term *identity-based cryptography* refers to cryptographic systems that derive public keys from user's identities. In 1984, Shamir introduced identity-based cryptography (IBC) (Shamir 1984). Shamir stated that:

> The scheme is based on a public key cryptosystem with an extra twist: instead of generating a random pair of public/secret keys and publishing one of these keys, the user chooses his name and network address as his public key. Any combination of name, social security number, street address, office number or telephone number can be used provided that it uniquely identifies the user in a way he cannot later deny, and that it is readily available to the other party. The corresponding secret key is computed by a PKG and issued to the user when he first joins the network.

He constructed an identity-based signature (IBS) scheme using the existing RSA function (Rivest et al. 1978), but was unsuccessful in constructing an identity-based encryption (IBE) scheme, which remained a long-standing problem for almost a decade. In 2001, Boneh and Franklin came up with an independent solution using the concept of bilinear maps (Boneh and Franklin 2001). This led to a new era of research in IBC, during which many identity-based digital signature schemes were proposed using the concept of bilinear maps. At the same time, Cocks proposed an IBE scheme using quadratic residuosity (Cocks 2001). However, Cocks' scheme was limited in its applicability to WSN due to the generation of long ciphers and slower performance, as it is based on a ternary quadratic form. Finally, Boneh, Lynn, and Shacham proposed the shortest-length identity-based digital signature scheme using the concept of pairing in classical cryptography (Boneh et al. 2004).

Identity-based cryptography steps up as a feasible solution against the large set of problems proposed by PKI. Needless to say, however, no crypto-system is perfect, and IBC has its associated costs, discussed later in this section. The basic idea behind IBE is that in any form of communication, the sender needs some basic information about the receiver (email address, phone number) before sending a message to the receiver.

As proposed by Shamir, it would greatly simplify secure communication if this basic information were transformed to a public key that could be used for encryption. Hence, IBE eliminates the whole process of distributing public key certificates as well as relaying revocation lists. From a serviceability point of view, this approach is extremely practical, as an encrypted message can be sent to anyone without worrying about the process required in retrieving the receiver's public key. On the other end, the receiver can use his private key corresponding to the public key to decrypt the message. Compared to the traditional PKI-based system, a *private key generator* (PKG) is responsible for distributing the private key. The PKG calculates the recipient's private key from his identity and sends it to him in a secure way. In a real environment, keys generated by the PKG are short-lived, and this feature of IBE introduces the concept of key freshness. Hence, recipients of an encrypted message are forced to retrieve their private keys from the PKG as soon as their key expires. In addition, the sender need not be concerned about the expiration time of the recipient's private key. If the receiver's identity is revoked, the PKG in his domain will stop issuing private keys to this user. Figure 4.8 outlines the IBE scheme. If Alice wants to send an encrypted message to Bob using the IBE scheme, she simply generates Bob's public key $[B(pu_k)]$ by hashing his identity $(I_B)$. Then she uses $B(pu_k)$ and encrypts the message $M$. At the receiver's end, Bob retrieves his private key $[B(pr_k)]$ from the PKG in his domain and decrypts the encrypted message $C$.

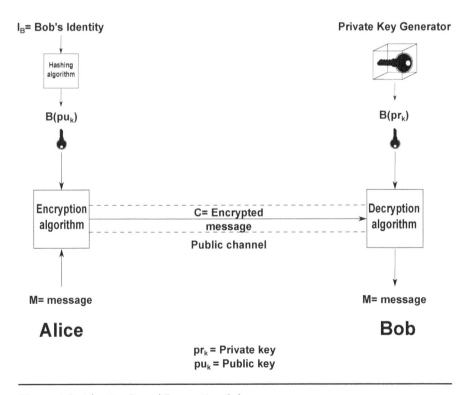

**Figure 4.8.** Identity-Based Encryption Scheme

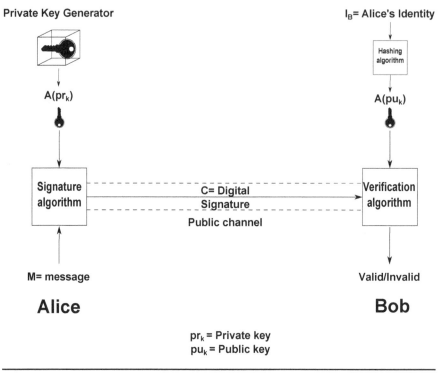

**Figure 4.9.** Identity-Based Signature Scheme

In an identity-based authentication scheme, Alice retrieves her private key $[A(pr_k)]$ from the PKG and the message $M$, and generates a digital signature using the identity-based digital signature algorithm (Figure 4.9). At the receiver's end, Bob simply hashes Alice's identity and generates her public key $[A(pu_k)]$ to verify the signature using the verification algorithm. In a hierarchical model, an identity-based authentication scheme significantly simplifies the verification process, as the recipient generates the public key from each of the intermediate PKG's identities. Compared to PKI, the recipient is forced to retrieve the public key certificate from each of the intermediate CAs and checks their validity before verifying the signature. In addition, IBC systems need not store any public key certificates, as public keys can be generated instantaneously as and when needed.

The security of IBC depends mainly on the secrecy of the information stored in the PKG, so considerable effort is involved in designing a secure PKG before deployment. In addition, care should be taken in validating the identities of end users before issuing their private key. One of the main advantages of this scheme is that the receiver need not have the private key corresponding to the public key at that very instant to decrypt the encrypted message. However, this approach has the inherent key escrow problem, which can actually be advantageous in certain applications such as WSN. In later sections, we will discuss this key escrow problem and ways to solve this problem. From a security standpoint, all users within a group or domain should trust their private key generators. Such a centralized trust model, however, could lead to a possibility of a single-point failure because of the PKG turning malicious. It could surreptitiously keep a copy of all the generated private keys and decrypt the encrypted data at a later time. This problem could be solved using the concept of threshold key cryptography, discussed later.

In a PKI-based system, each end user requires a public key certificate from all the intermediate CAs as well as the recipient's certificate to send an encrypted message. However, in an IBC environment, the problem of obtaining authentic public key certificates has been replaced with the task of obtaining authentic public parameters from the PKG. This is where IBC has a clear advantage, as there are fewer PKGs compared to end users. For instance, if there is a single PKG, all the end users can communicate securely with each other without ever querying for a public key certificate. In the next section, we review some of the computational problems that form the basis for the identity-based encryption and signature schemes.

### 4.3.1 Computational Problems

Computational problems are considered to be suitably hard, but can still be acknowledged in terms of quantities that provide cryptographic strength for public key algorithms. For instance, the Diffie-Hellman key exchange provides the motivation for many computational problems in cryptography. Let us review the classical Diffie-Hellman (D-H) key exchange protocol (Figure 4.10). Let Alice and Bob agree on an element g that generates the multiplicative group G. Then Alice and Bob select random values $r_A \in Z_q$ and $r_B \in Z_q$, and calculate $t_A = g^{r_A}$ and $t_B = g^{r_B}$, respectively. Given g, $t_B$, Alice computes the shared secret $Z_{AB} = t_B^{r_A}$. Similarly, Bob computes $Z_{AB} = t_A^{r_B}$ given g and $t_A$. Hence, g, $t_A$, and $t_B$ are publicly available, and without the private values $r_A$ and $r_B$, it is believed to be hard for an adversary to compute $Z_{AB}$. This general framework is used to discuss variants of the D-H key exchange in the subsequent sections. Multiplicative notation is used instead of additive notation because multiplicative

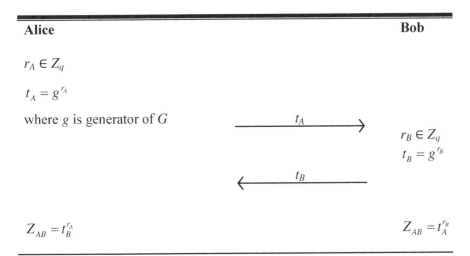

**Figure 4.10.** Diffie-Hellman Key Exchange Protocol

notation is usual for a D-H exchange. In addition, additive notation can also be used, where $P$ is the generator of the additive group.

In many situations, two related problems exist: a *computational* problem and a *decision* problem. Solving a *computational problem* is comparable to the process of calculating a correct answer; if the computational problem is hard, then calculating the correct answer is hard. A *decision problem* involves the decision of determining YES or NO as an answer to a problem. Here, if we want the correct answer to be difficult to guess or even determine part of the answer, and if the relevant decision problem is hard, then determining the correct answer is equally hard.

Let us look into computational and decision problems in the context of the Diffie-Hellman protocol.

### 4.3.1.1 Computational Diffie-Hellman Problem

The *computational Diffie-Hellman protocol* (CDHP) replicates the situation in D-H exchange. Given $g$, $t_A = g^{r_A}$, and $t_B = g^{r_B}$, calculate $Z_{AB} = g^{r_A r_B}$. Note that the CDHP can also be represented using additive notation: Given $P$, $r_A P$, $r_B P$, calculate $r_A r_B P$. For the sake of simplicity and maintaining regularity, let us use the multiplicative notation instead of additive notation. An adversary could take the obvious way to determine $r_B$ by calculating the discrete logarithm of $g^{r_B}$. Then he would calculate $Z_{AB}$ as shown below:

$$(t_A)^{r_B} = Z_{AB} \tag{4.1}$$

Hence, finding the discrete logarithm of either of the exchanged values would result in the discovery of the shared secret. However, solving the CDHP is no harder than solving the discrete logarithm problem. Despite considerable research in the area of computational complexity theory, the question of whether the D-H problem is as hard as the discrete logarithm problem still persists (Maurer 1994).

Conversely, given $g$, $t_A$, and $t_B$, there is no guarantee that an adversary cannot determine some part of the shared secret ($Z_{AB}$). (For instance, an adversary could determine several bits of $Z_{AB}$ but not all of the them.) Hence, to avoid such prediction from the adversary, another problem should be hard: the decision D-H problem.

### 4.3.1.2 Decision Diffie-Hellman Problem

The *decision Diffie-Hellman problem* (DDH) states that it is a hard problem to distinguish between a genuine D-H tuple $(t_A, t_B, Z_{AB})$ and a random tuple $(t_A', t_B', Z_{AB}')$, where $r_A'$, $r_B'$, and $z = r_A' r_B'$ are chosen randomly. One obvious but rudimentary way of solving this problem is to determine $r_B$ through the CDHP and then calculate $(t_A)^{r_B} = (g^{r_A})^{r_B} = Z_{AB}$ and compare $Z_{AB} = Z_{AB}'$. Hence solving the DDHP problem is no more difficult than solving the CDHP. If the DDHP is as hard as the CDHP, then it is equally hard to distinguish between $Z_{AB}$ and any other element in the group $G$.

### 4.3.1.3  Gap Diffie-Hellman Group

If $G$ is a cyclic group whose order is prime, it is said to be a *gap Diffie Hellman group* (GDHG) if the DDHP can be solved in polynomial time but no probabilistic algorithm can solve the CDHP with non-negligible advantage within the polynomial time (Okamoto and Pointcheval 2001).

### 4.3.1.4  Bilinear Diffie-Hellman Group

The *bilinear Diffie-Hellman group* (BDHG) generalizes the CDHP to groups with a pairing. This problem states that given $P$, $aP$, $bP$, $cP$, calculate $e(P,P)^{abc}$, where $a, b, c \in Z_q$. Note that an additive notation is used instead of a multiplicative notation, because the setting of the BDHP is typically an elliptic curve group. In the case of pairing, let us assume $e : G_1 \times G_2 \rightarrow G_T$. Solving the problem of bilinear Diffie-Hellman is no more difficult than calculating discrete logarithms in either $G_1$ or $G_T$. Hence, if we can find the value of $c$ by calculating the discrete logarithm of $cP$ in $G_1$, then we can calculate $e(aP, bP)^c = (e(P, P)^{ab})^c = e(P,P)^{abc}$, or we can find the value of $c$ by calculating the discrete logarithm of $e(P, cP) = e(P,P)^c$ in $G_2$. Then we can also calculate $e(P,P)^{abc}$ in a similar way.

## 4.3.2  Identity-Based Encryption Schemes

Four probabilistic polynomial time algorithms, Set-up, Key-Gen, Encrypt, and Decrypt, define identity-based encryption schemes.

> *Set-up*: This algorithm takes the security parameter $k$ and returns the master key $M_s$ and system parameters **PP**. **PP** includes a brief description of a finite message space **M** and cipher text **C**. Intuitively, the master secret key $M_s$ is generated by PKG while the **PP** is publicly known to end users. Let $id \in I$ (a set of identities).
>
> *Key-Gen*: This algorithm takes the user's identity id $\in I$ along with PP and the master secret $M_s$, and returns the corresponding user's private key did.
>
> *Encrypt*: This takes the identity $id \in I$, $m \in$ **M**, and public parameters **PP** and generates the cipher text $c \in$ **C**.
>
> *Decrypt*: This algorithm takes the identity $id \in I$, cipher text $c \in$ **C**, and a private key $d_{id}$ corresponding $id \in I$ and returns the message $m \in$ **M** or an error message.

Since the PKG generates the private keys corresponding to public keys, it can decrypt any message encrypted using the identity $id \in I$. This inherent property in identity-based cryptography is referred to as the *key escrow property*.

### 4.3.2.1 Boneh and Franklin IBE Scheme

Boneh and Franklin explain the basic idea of the identity-based encryption scheme through a simple scheme named *BasicIdent* (Boneh and Franklin 2001). This scheme uses a shared secret that can be calculated by both the sender and the receiver of a message to encrypt a plain text. Here we describe the basic scheme, which is secure against chosen plain-text attacks and adaptive chosen-identity attacks, while the full Boneh-Franklin IBE (Boneh and Franklin 2001) is resistant to chosen cipher-text attacks and adaptive chosen-identity attacks.

The *BasicIdent* scheme consists of *Set-up*, *Key-Gen*, *Encrypt*, and *Decrypt*.

*Set-up:* Let $k \in Z$ and the master secret $M_s \in Z_q^*$. Let $G$ be a bilinear Diffie-Hellman (BDH) parameter generator. Let $G_1$, $G_2$ be two groups of order $q$ and let $P \in G_2$ be a random generator. Let there exist a bilinear map $e$ such that

$$e : G_1 \times G_1 \rightarrow G_2 \qquad (4.2)$$

Let $P_{pub} = M_s P$.
Let the cryptographic hash functions be defined as

$$H_1 : \{0,1\}^* \rightarrow G_1^* \qquad (4.3)$$

$$H_2 : G_2 \rightarrow \{0,1\}^n \text{ for some } n \qquad (4.4)$$

Let the message space $M = \{0,1\}^n$ and the cipher text $C = G_1^* \times \{0,1\}^n$. Hence, the system parameters $PP = \langle G_1, G_2, q, e, n, H_1, H_2, P, P_{pub} \rangle$.

*Key-Gen:* The algorithm computes

$$Q_{id} = H_1(id) \in G_1^* \qquad (4.5)$$

where $id \in I$ and the private key corresponding to identity $id \in I$ is defined as

$$d_{id} = M_s Q_{id} \qquad (4.6)$$

*Encrypt:* This algorithm chooses a random integer $r \in Z_q^*$ and encrypts a message $m \in M$ to generate the cipher text

$$C = \langle U, V \rangle = \langle rP, m \oplus H_2(g_{id}^r) \rangle \text{ where } g_{id} = e(Q_{id}, P_{pub}) \in G_2^* \qquad (4.7)$$

*Decrypt:* Let $id$ be the public key used to generate the cipher text $C \in C$, where the private key $d_{id} \in G_1^*$, then the algorithm decrypts $C$ to generate $m \in M$ as shown below (Figure 4.11).

---

**Alice**                                                    **Bob**

$$PP = \langle G_1, G_2, q, e, n, H_1, H_2, P, P_{pub} \rangle$$

Let $m \in M$ and $r \in Z_q^*$

$$C = \langle U, V \rangle = \langle rP, m \oplus H_2(g_{id}^r) \rangle$$

where $g_{id} = e(Q_{id}, P_{pub}) \in G_2^*$

$$\xrightarrow{\quad\quad C \quad\quad}$$

$$V \oplus H_2(e(d_{ID}, U)) = m \in M$$

---

**Figure 4.11.** Boneh-Franklin IBE Basic Scheme

$$V \oplus H_2(e(d_{ID}, U)) = m \tag{4.8}$$

### 4.3.2.2 Boneh and Boyen IBE Scheme Without Random Oracles

Boneh and Boyen's IBE scheme belongs to the family of "commutative blinding" schemes (Boneh and Boyen 2004). Compared to the Boneh and Franklin IBE scheme, identity-based crypto-systems based on commutative blinding schemes hash map the identity to an integer rather than a point on the elliptic curve. Consequently, such schemes are faster than the full-domain hash schemes, as they avoid modular exponentiation. In this section, we discuss the basic Boneh-Boyen IBE scheme (Figure 4.12). This scheme consists of *Set-up*, *Key-Gen*, *Encrypt*, and *Decrypt*.

---

**Alice**                                                    **Bob**

$$PP = \langle G_1, G_T, q, e, H_3, H_4, P_1, P_2, P_3, v \rangle$$

$Q_{id} = H_3(id)$
$k = v^s \in G_T$
$c = m \oplus H_4(k)$ and
$C_0 = sP$ and
$C_1 = Q_{id}(sP_1) + sP_3$

$$\xrightarrow{\quad C = (c, C_0, C_1) \quad}$$

$$k = \frac{e(C_0, D_0)}{e(C_1, D_1)}$$

$$m = c \oplus H_4(k)$$

---

**Figure 4.12.** Boneh-Boyen Basic IBE Scheme

*Set-up*: Let $G_1$ be a subgroup of $E(F_q)$, where $E$ is an elliptic curve with an embedding degree $k$. Let $G_T$ be a subgroup of $F_{q^k}^*$, where $q$ is a prime power, and let there exist a bilinear map $e$ such that

$$e : G_1 \times G_1 \rightarrow G_T \tag{4.9}$$

Let $P \in G_1$ and let $\alpha, \beta, \gamma \in Z_q^*$. Let $P_1 = \alpha P$, $P_2 = \beta P$, and $P_3 = \gamma P$. Then there exists a constant $v \in G_1$ such that

$$v = e(P_1, P_2) = e(\alpha P, \beta P) = e(P, P)^{\alpha\beta} \tag{4.10}$$

Let $H_3$ be a hash function such that strings representing identities are mapped to integers,

$$H_3 : \{0,1\}^* \rightarrow Z_q \tag{4.11}$$

and let $H_4$ be another hash function such that

$$H_4 : G_T \rightarrow \{0,1\}^n \tag{4.12}$$

Hence, the system parameters $PP = \langle G_1, G_T, q, e, H_3, H_4, P_1, P_2, P_3, v \rangle$.

*Key-Gen*: Let $id \in I$, then this algorithm computes $Q_{id} = H_3(id) \in G_1$. Let $r \in Z_p$, then the private key is

$$d_{id} = \{Q_{id} \cdot rP_1 + \alpha P_2 + \gamma P_3, rP\} = \{D_0, D_1\} \tag{4.13}$$

This private key is sent to the end user with identity $id \in I$ in a secure way.

*Encrypt*: Let message $m \in M$, then this algorithm generates the cipher text as follows:

$Q_{id} = H_3(id)$.
Pick a random integer $s \in Z_p$ and calculate the constant $k = v^s \in G_T$.
Calculate $c = m \oplus H_4(k)$ and $C_0 = sP$.
Calculate $C_1 = Q_{id}(sP_1) + sP_3$.

Hence, the cipher text $C = (c, C_0, C_1)$.

*Decrypt*: This algorithm retrieves $k$ using the bilinear property as shown below:

$$k = \frac{e(C_0, D_0)}{e(C_1, D_1)} \tag{4.14}$$

Using $k$, the algorithm retrieves $m$.

$$m = c \oplus H_4(k) \tag{4.15}$$

Note: The Boneh-Boyen basic IBE scheme is vulnerable to chosen cipher-text attack.

### 4.3.3 Hierarchical Identity-Based Encryption Scheme

The computational cost of generating private keys can increase exponentially with the increase in the number of end users requesting private keys. Therefore, a single PKG is unfavorable in a large network. Furthermore, there is an additional cost of verifying the identities of end users and establishing secure channels for transmission of private keys. Hence, the motivation to adopt a *hierarchical identity-based encryption* (HIBE) scheme is even stronger, as it allows the root PKG to delegate work to its lower-level PKGs (Figure 4.13). The lower-level PKGs verify the authenticity of its users and distribute private keys within their specified boundaries.

Horwitz and Lynn (2002) first introduced HIBE, and a fully functional HIBE system was described by Gentry and Silverberg (2002). In this scheme, the root PKG generates private keys for its immediate lower-level PKGs, which in turn generate private keys for their lower-level PKGs or end users. Moreover, HIBE reduces the possibility of a total compromise of the network if any one of the intermediate PKG's private key is compromised.

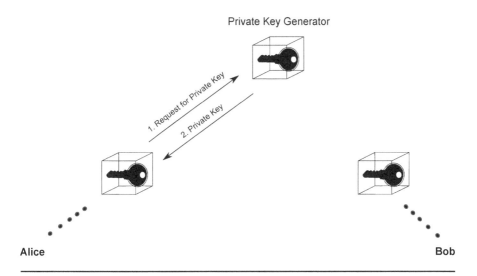

Private Key Generator

1. Request for Private Key

2. Private Key

Alice

Bob

**Figure 4.13.** Hierarchical Identity-Based System

In a hierarchical identity-based encryption scheme, a vector represents the identities. Let $h$ be the maximum height of the HIBE system. Analogous to the IBE scheme, the PKG sets the public parameters $PP$ and the master secret $M_s$.

Analogous to IBE, the HIBE scheme consists of *Set-up, Key-Gen, Encrypt,* and *Decrypt.*

*Set-up*: This takes the security parameters $k$ and returns the master private key and public parameters $PP$. The public parameters include the message space and the cipher-text space. In addition, the maximal level $h$ is also made public.

*Key-Gen*: This takes the identity-tuple $id = (id_1,\ldots, id_j)$ for entity $j$ and the private key $d_{id|j-1}$ for the identity $(id_1,\ldots, id_{j-1})$ and returns the private key $d_{id}$ using $d_{id|j-1}$.

*Encrypt*: This takes the identity $id \in I$, $m \in M$ and public parameters $PP$ and generates the cipher text $C \in C$

*Decrypt*: This algorithm takes the identity $id \in I$, cipher text $c \in C$, and a private key $d_{id}$ corresponding $id \in I$ and returns the message $m \in M$ or an error message.

### 4.3.3.1 Gentry-Silverberg HIBE

The Gentry and Silverberg (2002) HIBE is as follows.

*Set-up*: This scheme consists of $PP = \langle G_1, G_2, e, P_0, Q_0, H_1, H_2 \rangle$, where $G_1$ is an additive group of order $q$ and $G_2$ is a multiplicative group of order $q$. Let $e$ be a bilinear map such that $e : G_1 \times G_2 \to G_2$. Let $H_1$ and $H_2$ be cryptographic hash functions. Let $S_0$ be the identity element of $G_1$.

*Root Key-Gen*: Let root PKG pick $s_0 \in Z_q^*$ and generate $Q_0 = s_0 P_0$.

*Lower-Level Key-Gen*: Let entity $E_t$ at level $t$ (Level$_t$) have the ID-tuple $(ID_1,\ldots, ID_t)$. Let $E_t$'s ancestor be at Level$_i$, where $1 \le i \le t$. Let entity $E_t \in$ Level$_t$ pick a secret $s_t \in Z_q^*$.

*Extract*: $E_t$'s parent generates the following private key for $E_t$,

$$S_{t-1} + s_{t-1} P_t = \sum_{i=1}^{t} s_{i-1} P_i \qquad (4.16)$$

where $P_t = H_1(ID_1,\ldots, ID_t) \in G_1$. Furthermore, $E_t$'s parent distributes the value of $Q_i = s_i P_0$ for $1 \le i \le t-1$.

*Encrypt*: This algorithm encrypts the message $m \in M$ with the ID-tuple $(ID_1,\ldots, ID_t)$ as shown below:

$$C = \left\langle rP_0, rP_2, ..., rP_t, m \oplus H_2(g^r) \right\rangle \tag{4.17}$$

where

- $r \in Z_q^*$ is the secret picked by entity $t$ at Level$_t$, $P_i = H_1(ID_1, ..., ID_i) \in G_1$ for $1 \leq i \leq t-1$
- $g = e(Q_0, P_1) \in G_2$ $g = e(Q_0, P_1) \in G_2$

*Decrypt*: Let $C = \left\langle rP_0, rP_2, ..., rP_t, m \oplus H_2(g^r) \right\rangle = \left\langle U_0, U_2, ..., U_t, V \right\rangle$, where $c \in$ C. To decrypt $C$, the receiving entity at Level$_t$ computes

$$V \oplus H_2 \left( \frac{e(U_0, S_t)}{\displaystyle\prod_{t=2}^{t} e(Q_{i-1}, U_i)} \right) = m \in M \tag{4.18}$$

Although the HIBE is an extension of Boneh-Franklin IBE (discussed in Section 4.3.2.1) to a $t$-level IBE scheme, it still inherits the vulnerabilities of Boneh-Franklin IBE and is susceptible to chosen cipher-text attack.

### 4.3.3.2 Dual HIBE: Dual Hierarchical Identity-Based Encryption

The dual HIBE is structured as follows (Gentry and Silverberg 2002).

*Assumption*: End users $y$ and $z$ are close to each other in the hierarchy, with user $y$ at level $m$ (Level$_m$) and user $z$ at level $n$ (Level$_n$). Let there be a common ancestor at level $i$ (Level$_i$). Let user $y$ have ID-tuple $(ID_{y_1}, ..., ID_{y_l}, ..., ID_{y_m})$ and user $z$ have ID-tuple $(ID_{z_1}, ..., ID_{z_l}, ..., ID_{z_n})$ where $(ID_{y_1}, ..., ID_{y_l}) = (ID_{z_1}, ..., ID_{z_l})$.

*Encrypt*: User $y$ encrypts a message $m \in$ M as shown below:

$$C = \left\langle rP_0, rP_{z_{l+1}}, ..., rP_{z_n}, M \oplus H_2(g_{y_l}^r) \right\rangle \tag{4.19}$$

where

$$g_{y_l} = \frac{e(P_0, S_y)}{\displaystyle\prod_{i=l+1}^{m} e(Q_{y_{(i-1)}}, P_{y_i})} = e(P_0, S_{y_l})$$

$$P_{z_i} = H_1(ID_{z_1}, ..., ID_{z_i}) \in G_1 \text{ for } l+1 \leq i \leq n$$

$S_y$ is $y$'s secret and $S_{y_l}$ is the secret point of $y$'s and $z$'s common ancestor. $Q_{y_i} = s_{y_i} P_0$, where $s_{y_i}$ is the secret chosen by $y$'s ancestor at level $i$.

*Decrypt*: Let $C = \langle U_0, U_{l+1}, ..., U_n, V \rangle$ be the cipher text. User $z$ decrypts $C$ as shown below:

$$V \oplus H_2 \left| \frac{e(U_0, S_z)}{\prod_{i=l+1}^{n} e(Q_{z(i-1)}, U_i)} \right| = M \qquad (4.20)$$

## 4.3.4 Identity-Based Authentication Schemes

Digital signatures are one of the building blocks of cryptography that provide authentication, integrity, and nonrepudiation. The signer signs a message using his private key and a verifier verifies the signature using the signer's public key for the corresponding private key. In a traditional PKI-based architecture, the digital certificate binds the public key with its identity. However, Shamir's proposal eliminates the use of digital certificates by using the identity itself as a public key. Hence, ID-based signature schemes provide a simple architecture for verifying digital signatures.

### 4.3.4.1 BLS Short Signature Scheme

The BLS signature scheme (Boneh et al. 2004) makes use of a hash function $H_1$ and an asymmetric pairing $e : G_1 \times G_1 \rightarrow G_2$. Let $G_1 \in \langle P_1 \rangle$ and $G_2 \in \langle P_2 \rangle$.

*Key-Gen*: Computes $P_{pub} = xP_2$, where $x \in Z_p$ is the private key. Let $(P_2, P_{pub}) \in PP$.

*Signing*: Given a message $m \in M$, the signature $\sigma$ is calculated as

$$\sigma = xH_1(m) \in G_1 \qquad (4.21)$$

In this case, the signature is a single element in $G_1$.

*Verify*: Given a signature $\sigma$ and $Q = H_1(m)$, we verify the message $m \in M$ if the following condition holds:

$$e(\sigma, P_2) = e(Q, P_{pub}) \qquad (4.22)$$

If the above condition holds, the signature is valid; otherwise it is invalid.

## 4.3.4.2 ID-Based Blind Signature Scheme

*Set-up*: Let $G_1$, $G_2$ be two groups of order $q$, where $G_1$ is an additive group and $G_2$ is a multiplicative group. Let $e$ be the bilinear map that exists between $G_1$ and $G_2$ such that e: $G_1 \times G_2 \to G_2$. Let $H_1$ be the cryptographic hash function.

*Key-Gen*: The signer picks $x \in Z_q^*$ such that $P_0 = xP$, where $P \in G_1$.

*Signature*:

(*Blinding process*): Let $m \in M$ and the user picks $r \in Z_q^*$ and computes

$$M' = rH_1(m) \tag{4.23}$$

(*Signing process*): The signer simply signs the message $M'$ as shown below:

$$\sigma' = xM' \tag{4.24}$$

(*Unblinding process*): The user then computes the signature

$$\sigma = r^{-1}\sigma' \tag{4.25}$$

and sends $(m, \sigma)$.

*Verification*: Given the public parameter $P_0$ and the message $m$, the verification process is done as shown below:

$$e(P_0, H_1(m)) = e(P_0, \sigma) \tag{4.26}$$

## 4.3.4.3 Bilinear Ring Signature

The bilinear ring signature is structured as follows (Boneh et al. 2003).

*Set-up*: This signature scheme assumes a bilinear map and the Co-GDH set-up. Let $G_1$, $G_2$, and $G_T$ be multiplicative groups of order $q$ such that $e : G_1 \times G_2 \to G_T$ and having a computable isomorphism $\psi : G_1 \to G_2$. Let $H_3 : \{0,1\}^* \to G_2$.

Let there be $n$ users belonging to a set $U$ and each user has a unique public/private key pair. A ring signature is constructed using the public keys of all the $n$ users. However, the verifier can only conclude that any one of the $n$ users' private key was used in generating the digital signature. Hence, this property of the ring signature is known as *signer ambiguity*.

*Key-Gen*: Let $k \in Z_q^*$ be the secret and $v = g_1^k$ be the public parameter, where $G_1 = \langle g_1 \rangle$. Then let $G_2 = \langle g_2 \rangle$ and $g_2 = \psi(g_1)$.

*Signing*: Let $m \in M$ and $v_1, \ldots, v_n \in G_1$ be the public keys. Let $k$ be the private key corresponding to the public key $v_s$ for some $1 \le s \le n$. Let $x \in Z_q^*$ such that $\forall x \ne s$. The sender will compute

$$\sigma_s = \left( \frac{h}{\psi \left( \prod_{i \ne s} v_i^{a_i} \right)} \right)^{1/k} \tag{4.27}$$

where $h = H_3(m)$ and $\forall i \ne s$. Let $\sigma_i = g_2^{x_i}$; the output of the ring signature is $\sigma = (\sigma_1, \ldots, \sigma_n) \in G_2^n$

$$\sigma = (\sigma_1, \ldots, \sigma_n) \in G_2^n \tag{4.28}$$

*Verification*: Given the message $m \in M$, public keys $v_1, \ldots, v_n \in G_1$, and the ring signature $\sigma$, we compute $h = H_3(m)$ and verify as shown below:

$$e(g_1, h) = \prod_{i=1}^{n} e(v_i, \sigma_i) \tag{4.29}$$

## 4.3.4.4 Noninteractive Deniable Ring Authentication

The notation of noninteractive deniable ring authentication allows a signer to sign a message $m$ on behalf of an ad-hoc collection of participants, and to convince a designated verifier $V$ that this message is correct (Susilo and Mu 2003). Moreover, it is required that the designated verifier $V$ cannot convince any other third party that the message $m$ was indeed authenticated. For this purpose, chameleon hash functions are used. A *chameleon hash function* is associated with a pair of public and private keys and has the following properties.

1. Anyone who knows the public key can compute the associated hash function.
2. For people who do not have knowledge of the trapdoor (the secret key), the hash function is collision-resistant, i.e., it is infeasible to find two inputs which are mapped to the same output.
3. The trapdoor information holder can easily find collisions for every given input.

The following is an example of a chameleon hash based on the hardness of the discrete log problem.

$x \in Z_q^*$ is the private key of user $V$.

$y = g^x$ is the public key of user $V$.

For a given message $m \in Z_q^*$, choose a random $r \in Z_q^*$ and set

$$\text{Cham - Hash}_v(m,r) = g^m, y^r \tag{4.30}$$

Note that the above chameleon hash function is collision-resistant for any user $U \neq V$. User $V$ can always find any other message $\hat{m} \neq m$ and compute the appropriate $\hat{r}$ to find the same hash value, because he knows $x$ and can easily solve $m + xr = \hat{m} + x\hat{r}$.

*Key-Gen*: $x \in Z_q^*$ is the secret key and $v = g_1^x \in G_1$ is the public key of user $u$.

*Noninteractive Ring Signing*: Given public keys $v_1, \ldots, v_n \in G_1$, a message $m \in Z_q$, and a private key $x$ corresponding to one of the public keys $v_s$ for some $s$, do the following:

1. Choose randomly $r \in Z_q^*$ and compute $\hat{h} = H(\text{Cham-Hash}_v(m,r)) \in G_2$.
2. Choose randomly $a \in Z_q^* \; \forall i \neq s$.
3. Set

$$\sigma_s = \left( \frac{\hat{h}}{\psi\left(\prod_{i \neq s} v_i^{a_i}\right)} \right)^{1/x} \tag{4.31}$$

4. $\forall i \neq s$, set $\sigma_i = g_2^{a_i}$.
5. The ring signature is $\sigma = (\sigma_1, \ldots, \sigma_n) \in G_2^n$.

*Verification*: Given public keys $v_1, \ldots, v_n \in G_1$, a message $m \in M$, a ring signature $\sigma$ and $r$, compute

$$\hat{h} = H(\text{Cham - Hash}_V(m,r)) \tag{4.32}$$

and verify

$$e(g,h) = \prod_{i=1}^{n} e(v_i, \sigma_i) \tag{4.33}$$

## 4.3.4.5 Zhang-Safavi-Susilo (ZSS) Short Signature Scheme

The ZSS scheme is as follows (Zhang et al. 2004).

*Key-Gen*: Let $H_1$ be a hash function. The secret key is $x \in Z_q^*$ and the public key is $P_0 = xP$ for a signer.

*Sign*: Given a secret key $x$ and a message $m \in M$, compute the signature:

$$S = \left( \frac{1}{H_1(m) + x} \right) P \qquad (4.34)$$

*Verify*: Given a public key $P_{pub}$ and a message $m$, the verification of the signature is done as shown below:

$$e(H_1(m)P + P_{pub}, S) = e(P, P) \qquad (4.35)$$

### 4.3.4.6 ID-Based Blind Signature Scheme (Schnorr Type)

This signature scheme is as follows (Zhang and Kim 2002).

*Set-up*: Let $H_1: \{0,1\} \to G_1$ be a map-to-point hash function. Consider another hash function $H_1 : \{0,1\}^* \times G_2 \to Z_q$. Choose $s \in Z_q^*$ and set $P_{pub} = sP$. The master key is $s$ and the global public key is $P_{pub}$.

*Extract*: Given the signer's public identity, $ID \in \{0,1\}^*$, compute the public key $Q_{ID} = H_1(ID)$ and the private key $S_{ID} = sQ_{ID}$.

*Blind Signature Issuing Protocol*: Given a signer's private key $S_{ID}$ and a message $m \in \{0,1\}^*$.

*Initialization*: The signer randomly chooses a number $r \in Z_q$, computes $R = rP$, and sends $R$ to the user as a commitment.

*Blinding*: The user randomly chooses $a, b \in Z_q^*$ as blinding factors and computes $c = H(m, e(bQ_{ID} + R + aP, P_{pub})) c = H(m, e(bQ_{ID} + R + aP, P_{pub}))$.

*Signing*: The signer sends back $S$, where $S = cS_{ID} + rP_{pub}$.

*Unblinding*: The user computes $S' = S + aP_{pub}$ and $c' = c - b$ and outputs $(m, S', c')$. Then $(S', c')$ is the blind signature of the message $m$.

*Verification*: Accept if and only if $c' = H_1(m, e(S', P)e(Q_{ID}, P_{pub})^{-c'}$.

### 4.3.4.7 ID-Based Ring Signature

For the Zhang and Kim ID-based ring signature, the protocol is as follows (Zhang and Kim 2002).

*Set-up*: Let $H_1:\{0,1\}^* \to G_1$ be a map-to-point hash function and $H : \{0,1\}^* \to Z_q^*$ be another hash function. Choose $s \in Z_q$ and set $P_{pub} = sP$. The master key is $s$ and the global public key is $P_{pub}$.

*Extract*: Given public identity $ID \in \{0,1\}^*$, compute the public key $Q_{ID} = H_1(ID)$ and the secret $S_{ID} = sQ_{ID}$.

Let $ID_i$ be a user's identity and $S_{ID_i}$ be the private key associated with $ID_i$ for $i = 1,\ldots, n$. Let $L = \{ID_i : 1 \le i \le n\}$ be the set of identities. The real signer's identity $ID_k$ is listed in $L$.

*Signing*: Given signer's private key $S_{ID_k}$ and a message $m \in \mathbf{M}$

*Initialization*: Choose randomly an element $A \in G_1$ and compute $c_{k+1} = H(L\|m\|e(A, P))$.

*Forward ring sequence generation*: For $i = k + 1,\ldots, n - 1, 0, 1,\ldots, k - 1$, choose randomly $T_i \in G_1$ and compute $c_{i+1} = H(L\|m\|e(T_i, P)e(c_iH_1(ID_i), P_{pub})$

*Forming the ring*: Compute $T_k = A - c_k S_{ID_k}$.

*Output the ring signature*: The resulting signature for $m$ and $L$ is the $(n + 1)$-tuple $(c_0, T_0, T_1,\ldots, T_{n-1}) : (c_0, T_0, T_1,\ldots, T_{n-1})$.

*Verification*: Given $(c_0, T_0, T_1,\ldots, T_{n-1}, m$, and $L$, compute $c_{i+1} = H(L\|m\|e(T_i, P) e(c_iH_1(ID_i), P_{pub})$ for $i = 0, 1,\ldots, n - 1$. Accept if $c_n = c_0$ and reject otherwise.

### 4.3.4.8 Hess's Signature from Pairing

Hess's signature from pairing is done as follows (Hess 2003).

*Set-up*: Choose $s \in Z_q^*$ and set $P_{pub} = sP$. The master key is $s$ and the global public key is $P_{pub}$. Let $H_1:\{0,1\}^* \to G_1$ be a map-to-point hash function and $H : \{0,1\}^* \times G_2 \to Z_q^*$ be another hash function.

*Extract*: Given a public identity $ID \in \{0,1\}^*$, compute the public identity $Q_{ID} = H_1(ID)$ and the secret key $S_{ID} = sQ_{ID}$.

*Sign*: Given a secret key $S_{ID}$ and a message $m \in \{0,1\}^*$, the signer chooses an arbitrary $P_1 \in G_1$, random $k \in Z_q^*$, and computes

$$r = e(P_1, P)^k \tag{4.36}$$

$$v = H(m, r) \tag{4.37}$$

$$u = vS_{ID} + kP_1 \tag{4.38}$$

The signature is then the pair $(u, v) \in G \times Z_q^*$.

*Verify*: Given a public key $Q_{ID}$, a message $m$, and a signature $(u, v)$, the verifier computes

$$r = e(u, P)e(Q_{ID} - P_{pub})^{v} \qquad (4.39)$$

Accept the signature if and only if $v = H(m, r)$.

## 4.3.5 Key Distribution in IBC

### 4.3.5.1 Joux's One-Round, Three-Party Key Agreement Protocol

The protocol for Joux's one-round, three-party key agreement is as follows (Joux 2000).

> *Set-up*: Let $G_1$, $G_2$ be two groups of the same prime order $q$. Let $G_1$ be an additive group and $G_2$ be a multiplicative group. The public parameters are group $PP = (G_1, G_2, e, q, P)$.
>
> Consider $A$, $B$, $C$ with secret keys $a, b, c \in Z_q^*$, respectively, where
>
> $A$ sends $aP$ to both $B$, $C$.
> $B$ sends $bP$ to both $A$, $C$.
> $C$ sends $cP$ to both $A$, $B$.
> $A$ computes $K_A = e(bP, cP)^a$.
> $B$ computes $K_B = e(aP, cP)^b$.
> $C$ computes $K_C = e(aP, bP)^c$.
>
> The common agreed key of $A$, $B$, $C$ is

$$K_{ABC} = K_A = K_B = K_C = e(P, P)^{abc} \qquad (4.40)$$

## 4.3.6 Key Escrow Problem

The general key distribution problem refers to the onerous task of distributing secret keys between communicating parties to provide security properties such as authentication or confidentiality or both. However, with the advent of private key distribution, new problems arose, such as the inherent key escrow problem and the need for secure distribution of private keys (Lee et al. 2004). Lee et al. came up with a secure private key distribution algorithm in IBC. We briefly review this algorithm here.

Let $H_1$, $H_2$, and $H_3$ be three hash functions such that $H_1 : \{0, 1\}^l \rightarrow G_1$, where $l$ is the length of the plain text. Let $H_2 : \{0, 1\}^l \times G_2 \rightarrow Z_q$, where $Z_q = Z/qZ$ denotes integers mod $q$, where $q$ is a large prime. Therefore $Z_q$ denotes the group $\{0, 1, 2, \ldots, q - 1\}$ and $Z_q = Z/qZ$ .

Let $H_3 : G_2 \rightarrow Z_q^*$ . Let PKG specify two groups $G_1$ and $G_2$ of order $q$, where $G_1$ is an additive group and $G_2$ is a multiplicative group. Let $e$ be a bilinear map such that $e : G_1 \times G_1 \rightarrow G_2$ with the following properties:

- Bilinearity. Let $(x_1, x_2, \text{ and } y) \in G_1$. Then $e(x_1 + x_2, y) = e(x_1, y).e(x_2, y)$.
- Nondegeneracy. There exist $x \in G_1$ and $y \in G_1$ such that $e(x, y) \neq 1$.

In fact, $G_1$ is a point subgroup on an elliptic curve over a finite field and $G_2$ is a subgroup of a cyclic group of a larger finite field. The pairings are derived from the Weil-Tate or $\eta_T$ pairing. The PKG chooses a private key $s_0 \in Z_q$ and computes the master public key:

$$P_0 = P \qquad \text{where } P \in G_1 \qquad (4.41)$$

The security of the master public key is dependent on the elliptic curve discrete log problem. The PKG publishes the description of the groups $G_1$ and $G_2$, public key $P_0$, hash functions ($H_1$, $H_2$, and $H_3$), the bilinear map $e$, and the group element $P$. Alice and Bob choose their secrets to compute their blinding factors. Alice, with identity $ID_A$, chooses a random secret $x \in Z_q^*$ and computes a blinding factor $X = xP$. An eavesdropper will not be able to generate the private key, since he has no knowledge of the secret $x$. She then requests the PKG to issue a partial private key by sending $X$ and $ID_A$. The PKG will use some preshared credentials to verify the authenticity of an end user's identity.

The PKG validates Alice's identity ($ID_A$) and computes the public key of Alice as

$$Q_{ID_A} = H_1(ID_A) \qquad (4.42)$$

It computes a blinded partial private key as

$$Q_{bl_A} = H_3[e(s_0 X, P_0)]s_0 Q_{ID_A} \qquad (4.43)$$

It then generates a signature $Sig(Q_{bl_A})$ for integrity protection:

$$Sig(Q_{bl_A}) = s_0 Q_{bl_A} \qquad (4.44)$$

It sends $Sig(Q_{bl_A})$ and $Q_{bl_A}$ to Alice.
    Alice verifies the signature using the formula

$$e(Sig(Q_{bl_A}), P) \overset{?}{=} e(Q_{bl_A}, P_0) \qquad (4.45)$$

and finally retrieves her private key $D_{ID_A}$ by unblinding $Q_{bl_A}$ as follows:

$$D_{ID_A} = \frac{Q_{bl_A}}{H_3[e(P_0, P_0)^x]} \qquad (4.46)$$

Hence, there is a secure key exchange between the PKG and the end users.

### 4.3.7 Threshold Signature Scheme

*Key-Gen:* Let $G_1$, $G_2$ be multiplicative groups and $e : G_1 \times G_1 \to G_2$ is a bilinear map. Let $H : \{0,1\}^*$. Let there be $u_i$ servers, where $1 \le i \le n$. The private key $x \in Z_q^*$ is shared among the users using Shamir's secret sharing scheme such that any subset $S$ of $t + 1$ servers can reconstruct $x$ using Lagrange interpolation:

$$x = \sum_{i \in S} L_i x_i \tag{4.47}$$

where $x_i$ is the private key share and $y_i = g^{x_i}$ is the public key share of user $u_i$.

*Signature Share Generation:* To sign a message $m \in \{0,1\}^*$, user $u_i$ is behaving honestly in giving its share $\sigma_i$ of the signature by checking

$$e(g, \sigma_i) = e(y_i, H(m)) \tag{4.48}$$

If $\sigma_i$ passes this test, call it an acceptable share.

*Signature Reconstruction:* Suppose a set $S$ of $(t + 1)$ honest servers is found and accordingly $(t + 1)$ acceptable shares $\sigma$, $i \in S$. The resulting signature on $m$ is $\sigma = \prod_{i \in S} \sigma_i^{L_i}$. The correctness of the scheme is easy to verify since

$$e(g, \sigma) = e(H(m), g^x) \tag{4.49}$$

#### 4.3.7.1 Pairing-Based (t, n) Threshold Decryption

The pairing-based $(t, n)$ threshold decryption scheme (Libert and Quisquater 2003) is a threshold adaptation of the Boneh-Franklin IBE scheme in which a fixed PKG plays the role of a trusted dealer.

*Set-up:* $PP = (G_1, G_2, q, e, H_1, H_2, P, P_{pub}^{(1)}, \ldots, P_{pub}^{(n)}, P_{pub})$.

1. Generate two groups $G_1$ (additive) and $G_2$ (multiplicative) of the same prime order $q$, $e : G_1 \times G_1 \to G_2$ is a bilinear map, $G_1$ is generated by $P$, $s \in Z_q^*$ is the master key.
2. Choose a $(t \times 1)$-degree polynomial.

$$f(x) = s + a_1 x + \cdots + a_{t-1} x^{t-1} \tag{4.50}$$

for a random $a_1, \ldots, a_{t-1} \in Z_q^*$

3. For $i = 1, 2,\ldots, n$, compute $P^{(i)}_{pub} = f(i)P \in G_1$ and $P_{pub} = sP$.

Before requesting his private share, each player can check that

$$\sum_{i \in S} L_i P^{(i)}_{pub} = P_{pub} \qquad (4.51)$$

for any subset $S \subset \{1,\ldots,n\}$ such that $|S| = t$, where $L_i$ denotes the appropriate Lagrange coefficient.

*Key-Gen*: Given a user's identity *ID*, the PKG plays the role of the trusted dealer. For $i = 1,\ldots, n$ it delivers $d_{ID_i} = f(i)Q_{ID} \in G_1$ of player $i$. After receiving $d_{ID}$, player $i$ checks

$$e(P^{(i)}_{pub}, Q_{ID}) = e(P, d_{ID_i}) \qquad (4.52)$$

If verification fails, he complains to the PKG, which issues a new share.

*Encrypt*: Given messages $m$ and identity ID,

1. Compute $Q_{ID} = H_1(ID)$.
2. Choose a random $r \in Z_q^*$.
3. Set the cipher text to be

$$C = \left\langle rP, m \oplus H_2(e(P_{pub}, Q_{ID})^r \right\rangle \qquad (4.53)$$

*Decrypt*: Upon receiving $\langle U, V \rangle$, player $i$ computes his decryption share $e(U, d_{ID_i})$ and gives it to the recombiner, who may be a designated player.

*Recombination*: The recombiner selects a set $S \subset \{1,\ldots,n\}$ of $t$ acceptable shares $e(U, d_{ID_i})$ and computes

$$g = \prod_{i \in S} e(U, d_{ID_i})^{l_i} \qquad (4.54)$$

Once he has $g$, he recovers the plain text

$$m = V \oplus H_2(g) \qquad (4.55)$$

The correctness of the scheme is easy to verify since

$$g = e\left(rP, \sum_{i \in S} L_i, d_{ID_i}\right) = e(rP, sQ_{ID}) = e(P_{pub}, Q_{ID})^r \qquad (4.56)$$

To check publicly whether the share of a player is acceptable, each player chooses a random $R \in G_1$ and computes $w_1 = e(P, R)$, $w_2 = e(U, R)$ and

$$h = H(e(U, d_{ID_i}), e(P_{pub}, Q_{ID}), w_1, w_2) \tag{4.57}$$

Then player $i$ computes $V = R + h d_{ID_i} \in G_1$ and joins the tuple $(w_1, w_2, h, V)$ to its share. The other players can check that

$$e(P, V) = e(P, R)e(P_{pub}^{(i)}, Q_{ID})^h \tag{4.58}$$

$$e(U, V) = e(U, R)e(U, d_{ID_i})^h \tag{4.59}$$

If this test fails, player $i$ is a dishonest player.

### 4.3.7.2 ID-Based (t, n) Threshold Decryption

The ID-based $(t, n)$ threshold decryption of Baek and Zheng (2004) works as follows.

*Key-Gen*: $PP = (G_1, G_2, q, e, P, H_1, H_2, H_3, H_4, P_{pub})$, where $P_{pub} = sP$ and $x \in Z_q^*$ is the master key of PKG.

*Extract*: Given an identity $ID$, compute $Q_{ID} = H_3(ID)$, $D_{ID} = xQ_{ID}$ and return $D_{ID}$.

*Private Key Distribution*: Given a private key $D_{ID}$, $n$ decryption shares and a threshold parameter, $t \leq n$, pick randomly $R_1, R_2, ..., R_{t-1} \in G_1^*$ and compute

$$F(u) = D_{ID} + uR_1 + u^2 R_2 + \cdots + u_{t-1} R_{t-1} \tag{4.60}$$

for $u \in \{0\} \cup N$. Compute $S_i = F(i)$, $y_i = e(S_i, P)$ $y_i = e(S_i, P)$, $1 \leq i \leq n$, and send $(S_i, y_i)$ secretly to server $T_i$, $1 \leq i \leq n$. $T_i$ then keeps $S_i$ secret while it publishes $y_i$.

*Encrypt*: Given a plain text $m \in \{0, 1\}^l$, identity $ID$,

1. Choose $r \in Z_q^*$ at random and set $U = rP$.
2. Compute $Q_{ID} = H_3(ID)$, $d = e(Q_{ID}, P_{pub})$, and $k = d^r$.
3. Compute $V = H_1(k) \oplus m$, $W = rH_2(U, V)$, $W = rH_2(U, V)$.
4. Set the cipher text to be $C = (U, V, W)$.

*Decrypt Share Generation*: Given a cipher text $C = (U, V, W)$, decryption server $T_i$, with secret key $S_i$, compute $H_2 = H_2(U, V)$ and check if $e(P, W) = e(U, H_2)$. If the test holds, then compute

$$k_i = e(S_i, U), \tilde{k}_i = e(Q_i, U), \tilde{y}_i = e(Q_i, P), \lambda_i = H_4(k_i, \tilde{k}_i, \tilde{y}_i) \text{ and } L_i = Q_i + \lambda S_i \tag{4.61}$$

*Decrypt Share Verification*: Given a cipher text $C = (U, V, W)$ and a decryption share $\delta_i = (i, k_i, \tilde{k}_i, \tilde{y}_i, L_i)$, compute $\lambda_i = H_4(k_i, \tilde{k}_i, \tilde{y}_i)$. Check if

$$\frac{e(L_i, U)}{k_i^{\lambda_i}} = \tilde{k}_i, \; \frac{e(L_i, P)}{y_i^{\lambda_i}} = \tilde{y}_i \qquad (4.62)$$

If the above test holds, then $\delta_i$ of server $T_i$ is an acceptable share. Given acceptable shares $S_j, j \in S \subseteq \{1, ..., n\}$, where $|S| \geq t$, $D_{ID}$ can be recovered as follows:

$$D_{ID} = F(0) = \sum_{j=S} c_{0j} S_j \qquad (4.63)$$

where the $c_{0j}$ are appropriate Lagrange coefficients.

*Share Combining*: Given a cipher text $C = (U, V, W)$ and a set of decryption shares $\{\delta_j\}_{j \in S \subseteq \{1,2,...,n\}}$, where $|S| \geq t$, compute $H_2 = H_2(U, V)$ and check whether $e(P, W) = e(U, H_2)$. If $C$ passes this test (i.e., $C$ is a valid cipher text), compute $k = \prod_{j \in S} k_j^{c_{0j}}$ and $m = H_1(k) \oplus V$. Output $m$.

The correctness of the scheme is easy to verify since

$$\prod_{j \in S} k_j^{c_{0j}} = \prod_{j \in S} e(S_j, U)^{c_{0j}} = e\left(\sum_{j \in S} c_{0j} S_j, U\right) = e\left(\sum_{j \in S} c_{0j} S_j, rP\right) = e(D_{ID}, P)^r \quad (4.64)$$

### 4.3.7.3 Sakai-Ohgishi-Kasahara Key Sharing Scheme

The idea of the Sakai-Ohgishi-Kasahara key sharing scheme (Sakai et al. 2000) is quite simple. Suppose a PKG has a master key $s$, and it issues private keys to users of the form $sP_y$, where $P_y = H_1(ID_y)$ and $ID_y$ is the identity of user $y$. Then users $y$ and $z$ have a shared secret that only they may compute:

$$e(sP_y, P_z) = e(P_y, P_z)^s = e(P_y, sP_z) \qquad (4.65)$$

They may use this shared secret to encrypt their communications. This key sharing scheme is noninteractive and can be viewed as a type of "dual-identity-based encryption," where the word "dual" indicates that the identities of both the sender and the recipient are required as input into the encryption and decryption algorithms.

## 4.4 Summary

This chapter has introduced public key cryptography and looked at different architectures of public key infrastructure (PKI). We then presented ID-based cryptography and provided a comparative analysis between ID-based cryptography and PKI. We reviewed different ID encryption, signature, and signcryption schemes in single and hierarchical IBE. Finally, ID-based threshold key cryptography has also been discussed.

## 4.5 References

Baek, J., and Y. Zheng. Identity-based threshold decryption. *Lecture Notes in Computer Science* (Springer-Verlag) 2947 (2004): 262–276.

Boneh, D., and X. Boyen. Efficient selective-ID secure identity-based encryption without random oracles. *Lecture Notes in Computer Science* (Springer-Verlag) 3027 (2004): 223–238.

Boneh, D., and M. Franklin. Identity-based encryption from the Weil pairing. *Lecture Notes in Computer Science* (Springer-Verlag) 2139 (2001): 213–229.

Boneh, D., C. Gentry, B. Lynn, and H. Shacham. Aggregate and verifiably encrypted signatures from bilinear maps. *Lecture Notes in Computer Science* (Springer-Verlag) 2656 (2003): 416–432.

Boneh, D., B. Lynn, and H. Shacham. Short signatures from the Weil pairing. *Journal of Cryptology* (Springer-Verlag), 2004: 297–319.

Cocks, C. An identity based encryption scheme based on quadratic residues. *Cryptography and in Coding—Institute of Mathematics and Its Applications International Conference on Cryptography and Coding—Proceedings of IMA 2001, LNCS* (Springer-Verlag) 2260 (2001): 360–363.

Diffie, W., and M. Hellman. New directions in cryptography. *IEEE Transactions of Information Theory* 22 (1976): 644–654.

Elgamal, T. A public key cryptosystem and a signature scheme based on discrete logarithms. *Transactions on Information Theory* 31, no. 4 (1985): 469–472.

Gentry, C., and A. Silverberg. Hierarchical ID-based cryptography. *Lecture Notes in Computer Science* (Springer-Verlag) 2501 (2002): 548–566.

Hess, F. Efficient identity based signature schemes based on pairings. *Selected Areas in Cryptography: 9th Annual International Workshop (Lecture Notes in Computer Science)* (Springer-Verlag) 2595 (2003): 310–324.

Horwitz, J., and B. Lynn. Toward hierarchical identity-based encryption. *EUROCRYPT'02, Lecture Notes in Computer Science* (Springer) 2332 (2002): 466–481.

Joux, A. A one round protocol for tripartite Diffie–Hellman. *Lecture Notes in Computer Science* (Springer-Verlag) 1838 (2000): 385–394.

Libert, B., and J. Quisquater. Efficient revocation and threshold pairing based cryptosystems. *Proceedings of the Twenty-Second Annual Symposium on Principles of Distributed Computing.* ACM (2003): 163–171.

Lee, B., C. Boyd, E. Dawson, K. Kim, and J. Yang. Secure key issuing in ID-based cryptography. *Conferences in Research and Practice in Information Technology* (2004): 69–74.

Maurer, U. M. Towards the equivalence of breaking the Diffie-Hellman protocol and computing discrete logarithms. *Lecture Notes in Computer Science* (Springer) 839 (1994): 271–281.

Okamoto, T., and D. Pointcheval. The gap-problems: A new class of problems for the security of cryptographic. *Lecture Notes in Computer Sciences* (Springer-Verlag) 1992 (2001): 104–118.

Rivest, R., A. Shamir, and L. Adleman. A method of obtaining digital signatures and public-key cryptosystems. *Communications* (ACM) 21, no. 2 (1978): 120–126.

Sakai, R., K. Ohgishi, and M. Kasahara. Cryptosystems based on pairing. *The 2000 Symposium on Cryptography and Information Security* (2000).

Shamir, A. Identity-based cryptosystems and signature schemes. *Advances in Cryptology: Proceedings of CRYPTO 84, Lecture Notes in Computer Science* (Springer-Verlag) 196 (1984): 47–53.

Susilo, W., and Y. Mu. Non-interactive deniable ring authentication. *Lecture Notes in Computer Science* (Springer-Verlag) 2971 (2003): 386–401.

Zhang, F., and K. Kim. ID-based blind signature and ring signature from pairings. *Lecture Notes of Computer Science* (Springer-Verlag) 2501 (2002): 629–637.

Zhang, F., R. Safavi-Naini, and W. Susilo. An efficient signature scheme from bilinear pairings and its applications. *Lecture Notes in Computer Science* (Springer-Verlag) 2947 (2004): 277–290.

# Chapter 5

# Identity-Based Authentication and Encryption in Wireless Sensor Networks*

*Large portions of this chapter were originally published in the following papers:

Rehana Yasmin, Eike Ritter, and Guilin Wang, An Authentication Framework for Wireless Sensor Networks using Identity-Based Signatures, *2010 10th IEEE International Conference on Computer and Information Technology*, 2010, 882–889.

Rehana Yasmin, An Authentication Framework for Wireless Sensor Networks Using Identity-Based Signatures: Implementation and Evaluation, Special section on Trust, Security and Privacy in Computing and Communication Systems in *IEICE Transactions on Information and Systems (IEICE-2011)*. Vol. E95-D, No. 01, 126–133, January 1, 2012.

Rehana Yasmin, Eike Ritter, and, Guilin Wang, Pairing-Free ID-based One-Pass Authenticated Key Establishment Protocol for Wireless Sensor Networks, presented at the *Fifth International Conference on Sensor Technologies and Applications* from August 21–27, 2011.

## In This Chapter

- Introduction to authentication in WSN
- Review of authentication schemes in WSN
- Identity-based authentication in WSN
- Identity-based encryption in WSN
- One-pass authenticated session key establishment in WSN
- Revocation in WSN

## 5.1 Introduction

Wireless Sensor Networks (WSN) consist of a large number of low-cost, low-powered sensor nodes that facilitate data collection in many diverse applications. They offer unprecedented opportunities for a broad range of monitoring and tracking applications in both civilian and military sectors (Chapter 1). In addition, the data aggregated from the sensor nodes are accessed from nodes within the network as well as external users. In all these applications, authentication and further encryption are fundamental security requirements and are useful in mitigating impersonation attacks. They are also useful in preventing the ever-increasing DoS and DDoS attacks on limited resource constraint environments such as WSN. Three scenarios exist in WSN that require authenticated communication:

- Sink node to sensor nodes and vice versa
- Sensor node with other sensor nodes
- Outside user and sensor nodes

Most of the time, critical applications in WSN require a message to be sent as promptly as possible. The intermediate nodes between the sender and receiver are responsible for relaying the message to the receiver. If one of the nodes is compromised, the malicious node can inject falsified packets into the network while routing messages. Such an act could lead to falsified distribution of such messages and, in turn, deplete the energy levels of other honest nodes. Hence, there is a need to filter messages as early as possible by authenticating every message, consequently conserving relaying energy.

In most WSN applications, the sensor nodes are expected to aggregate, process, store, and supply sensed data upon the end-user's query. For example, in a military application, soldiers would require constant interaction with motion sensors that detect any movement along the border. In such situations, a large number of mobile or static end users could query the sensor nodes for sensed data. Usually, such interactions are realized through broadcast/multicast operations. Therefore, in such situations, a broadcast authentication mechanism is required before the query is sent. Furthermore, access control is also required, which would only allow the authorized user to access data to which he is entitled. Broadcast authentication was first addressed by µTimed Efficient Stream Loss-tolerant Authentication (µTESLA) (Perrig et al. 2001). In this scheme, users are assumed to be a few trustworthy sink nodes. This scheme uses one-way hash functions and the hash pre-images are used as keys to the Message Authentication Code (MAC) algorithm.

However, the messages are transmitted through a wireless medium, which consumes a considerable amount of time. In addition, the hop-by-hop routing nature of WSN further creates a delay in transmission. Hence, there is an increased need for rapid generation and verification of signature schemes.

The existing symmetric schemes such as µTESLA and its variants use Message Authentication Code (MAC) to gain efficiency in terms of processing and energy consumption. However, these symmetric schemes suffer from delayed authentication and sluggish performance for large-scale networks, and they are susceptible to DoS attacks due to late authentication. Furthermore, multiple senders cannot send authenticated broadcast messages simultaneously. For example, if a single node is interested in broadcasting a message, it would have to send a Unicast message to its respective sink node, which would then broadcast the message to all the other nodes on its behalf. Because of resource constrains, asymmetric schemes—for example, digital signatures that would require public key certificates—were pronounced inefficient. Hence, to address this problem, security researchers and cryptographers started exploring new avenues to introduce authentication in public-key cryptography in WSN. One of the approaches proposed to address authenticated broadcast/multicast uses identity-based authentication. Using this scheme, two schemes are proposed, namely, Identity-Based Signature (IBS) and Identity-Based Online/Offline Signature (IBOOS). IBOOS is the ID-based version of Online/Offline Signature (OOS). For IBS, the public key is a unique identity associated to an entity, and the corresponding private key is generated by the private key generator (PKG) (Chapter 4). The receiver simply verifies the ID-based digital signature using the sender's identity. The need for a public-key certificate is totally eliminated in this scheme. The Online/Offline Signature schemes divide the message signing process into two phases, namely, Offline phase and Online phase. The Offline phase comes into action before the message to be signed becomes available and results in a partial signature. This phase is responsible for most of the precomputations. As soon as the message to be signed becomes available, the Online phase becomes active. In this phase, the partial signature generated in the Offline phase is used to generate the final signature. The idea behind using two phases instead of one is that the Offline phase could be performed by a resource abundant device (sink node) while the Online phase could be executed by the sensor nodes.

## 5.2 Related Work

### 5.2.1 *SPINS*

SPINS is a set of security protocols optimized for sensor networks and is built upon two secure building blocks, namely, Secure Network Encryption Protocol (SNEP) and μTESLA (Perrig et al. 2001). SNEP supports authentication, integrity, freshness, and data confidentiality, whereas μTESLA only supports broadcast data authentication and this operation is performed by the sink node. It uses a hierarchical architecture with a forest-like network formed around one or more sink nodes. Concerning key distribution, a unique master key is shared between the nodes and the base station. This master key is preloaded in each node before deployment. SNEP facilitates the establishment of the session key via the sink node or the base station. Two-party authentication and data integrity are achieved using MAC, which enforces message ordering and weak freshness. Concerning encryption, SNEP aims to achieve semantic security by sharing a counter between the sender and receiver for the block cipher in a counter mode. μTESLA uses a one-way key chain mechanism whereby the sink node randomly chooses a key and generates the remaining values by successively applying a one-way hash function.

#### 5.2.1.1 SNEP

In SNEP, all cryptographic operations such as encryption, decryption, pseudo random generation, and hash operations use the RC5 function. Furthermore, SNEP derives independent keys for encryption and MAC operations. To provide semantic security, a counter $C_A$ is used like an Initialization Vector (IV). The counter value is sufficiently long enough such that it is never repeated during a node's lifetime. However, in a resource constraint environment such as WSN, adding a counter with each message could severely increase storage and communication overhead. Hence, to avoid this issue, the counter is shared between the sender and the receiver. Since the state of the counter is kept at each end of the communicating party, the need for repeated transmission can be completely avoided.

SNEP uses the Cipher Block Chaining-Message Authentication Code (CBC-MAC) scheme to construct MAC. The counter value is included during the calculation of MAC to prevent any form of replay attacks. For instance, if A is interested in talking to B securely, A would construct a message $\{D\}_{\langle K_{AB},C_A \rangle}$ and $MAC_{K'_{AB}}[C_A \| \{D\}_{\langle K_{AB},C_A \rangle}]$ , where $K_{AB}$ is the encryption key that is derived from the master secret $X_{AB}$, which is used to encrypt data D. $K'_{AB}$ is a MAC key that is derived from $X_{AB}$. In addition, strong freshness that includes delay estimation could be achieved by considering a nonce N value during MAC calculations. Then, A would send the following to B: $\{D\}_{\langle K_{AB},C_A \rangle}$ and $MAC_{K'_{AB}}[N_A \| C_A \| \{D\}_{\langle K_{AB},C_A \rangle}]$. Although the counter value is not being sent with the transmitted packet, in a real environment packets could possibly be dropped during transmission. This could lead to a synchronization problem. Protocols that would need resynchronization would simply bootstrap the counter value. For example, A would send the counter value $C_A$ to B. B would respond with $C_B$ and $MAC_{K'_{BA}}[C_B,C_A]$. A would send back $MAC_{K'_{AB}}[C_A \| C_B]$.

### 5.2.1.2 μTESLA

The main focus of μTESLA is to provide authentication for broadcasted data. Primarily, it adopts the delayed key disclosure mechanism proposed in TESLA (Perrig et al. 2000). In this mechanism, the key utilized to authenticate the $i$th message is disclosed during the $(i + 1)$th message. The sink node randomly selects the last key $(K_n)$ of a chain and generates the rest of the keys $(K_0, K_1, K_2, \ldots, K_{n-1})$ in the chain as $K_i = h(k_{i+1})$. As a result, if a node is given $K_i$, it is able to generate keys $(K_0, K_1, K_2, \ldots, K_{i-1})$. However, this node cannot generate $K_{i+1}$. Upon deployment of the sensor nodes, the sink node sends an authenticated message $MAC_{K_i}[D]$ during the time slot and the leaf nodes are required to store this message until the sink node sends the verification key $(K_i)$ at the l + $d$th time slot. The parameter $d$ is the delay in the disclosure. Subsequently, the leaf nodes use the current disclosed key $K_i$ to verify the previous key: $K_{i-1} = h(k_i)$. However, this delayed key disclosure mechanism would require the storage of keys until the authentication key is received. This mechanism could lead to storage problems, consequently leading to DoS attacks. For instance, the adversary could flood nodes with fake keys that would exhaust the storage capacity of each node. As in SNEP, μTESLA incorporates the bootstrapping procedure to overcome the synchronization problem. However, bootstrapping would demand unicast communication, which would result in an increased volume of message flow in the network and would result in network congestion with the increase in scalability. Consequently, the option of multiple sender nodes broadcasting simultaneously is totally ruled out. To overcome this problem, μTESLA extensions have been proposed to address scalability issues by including two or more key chain levels (Liu and Ning 2003).

## 5.2.2 TinySec

TinySec is a lightweight link-layer security architecture for WSN (Karlof et al. 2004). It was designed with the intention of being easily portable on most of the existing sensors and addresses message integrity, data confidentiality, replay protection, and access control. However, it does not address attacks related to resource exhaustion and node capture.

This security framework supports two modes of architecture, namely, authenticated encryption *(TinySec-AE)* and authentication only *(TinySec-Auth)*. *TinySec-AE* is a full-security suite and offers packet authentication through Message Authentication Code (MAC) and data payload encryption, whereas in *TinySec-Auth* MAC is used to authenticate the entire packet, but the payload is not encrypted. MAC can be viewed as a distinct class of hash functions that facilitates message authentication and integrity through symmetric techniques. The intent of designing a MAC is that it is computationally infeasible to produce a matching output without prior knowledge of the key. The sender sends the plain text and its corresponding MAC to the receiver. The receiver ensures message authenticity and integrity by recomputing MAC and comparing it with the sender's MAC. MAC is also useful in detecting transmission errors so that the Bit Error Rate (BER) is not affected.

In TinySec, Cipher Block Chaining (CBC) is used for encryption with an 8-byte Initialization Vector (IV) to achieve semantic security, i.e., encrypting the plain text two times should result in different cipher texts. The length of the IV plays a crucial role in security and performance in any network. Especially in a WSN environment, if the length of the IV is too long, there would be an unnecessary addition of redundant data that would result in increased usage of expensive resources and degradation of overall throughput. In contrast, if the IV is too short, there could be the risk of a security attack due to the possibility of a repeating IV. Using the pigeonhole principle mentioned in Karlof et al. (2004), an IV in this model encompasses a counter (CTR) to prevent repetitions of IVs. The structure of an IV includes Destination address (DEST), Active Message (AM) handler type, Length of data payload (LEN), Source address (SRC), and a 16-bit counter. AM in this model carries the same functionality as port numbers in the TCP/IP model.

Figures 5.1 and 5.2 depict the packet formats in bytes for *TinySec-AE* and *TinySec-Auth*, respectively. Although this model demonstrates the ability to effectively

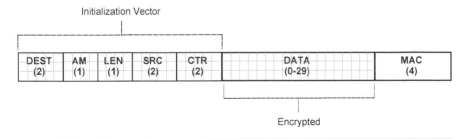

**Figure 5.1.** *TinySec-AE* Packet Format in Bytes

| DEST (2) | AM (1) | LEN (1) | | | | DATA (0-29) | | | | | MAC (4) |
|---|---|---|---|---|---|---|---|---|---|---|---|

**Figure 5.2.** *TinySec-Auth* Packet Format in Bytes

incorporate a link-layer cryptographic mechanism without any hardware support, it does not address the key-distribution problem (Chapter 6).

## 5.2.3 Lightweight Public Key Infrastructure for WSN

Although the applicability of Public Key Infrastructure (PKI)-based approaches (Section 4.2) has been deemed inappropriate for a resource constraint environment such as WSN, security researchers have been proposing new lightweight PKI-based approaches for WSN. For instance, a simplified version of Secure Socket Layer (SSL) has been proposed in WSN (Wander et al. 2005). Although this SSL version has a lesser overhead when compared to the usual SSL/TLS protocol, it is still not directly applicable to mobile sensor nodes because it would lead to increased communication and computational overhead. For instance, in an ad-hoc mobile sensor network, the nodes keep changing their location, and any change in their position would compel them to initiate the SSL protocol before informing their neighbors of their new location. In addition, schemes such as TinyPK have been designed that are in conjunction with TinySec and facilitate authentication and key agreement between sensor nodes (Watro et al. 2004). However, TinyPK implements Diffie-Hellman key exchange protocol that is susceptible to a man-in-the-middle attack. Huang et al. (2003) proposed a hybrid architecture for authenticated key establishment of a session key between a leaf node and a sink node or an end user. This protocol leverages on the difference in the computational and communication capabilities between the leaf node and the resource abundant device (sink node or end user). During the inception of the protocol, both parties exchange certificates issued by a Certificate Authority (CA) to extract each other's public keys. However, the corresponding private keys are discovered after both parties run the protocol. This step in this protocol can easily be exploited by an adversary by replaying a valid certificate that would result in a DoS attack. As a result, the nodes are forced to perform expensive computations and waste their resources and bandwidth. In addition, Tian et al. (2005) showed a serious vulnerability in Huang et al.'s scheme wherein an end user can easily discover the long-term private key of a leaf node after having one normal run of the protocol.

To expunge the transmission of public key certificates, Ren et al. (2012) propose a Hybrid Authentication Scheme (HAS) for a multiuser broadcast authentication scheme in WSN. In this scheme, each sensor node is preloaded with the required public key information of the end user using the Bloom filter and Merkle hash tree (Merkle 1980; Mitzenmacher 2002). However, HAS with the Merkle hash tree does not facilitate user scalability, i.e., a new user can only be added into the network after revocation of the old user.

## 5.3 Identity-Based Digital Signature Schemes in WSN

### 5.3.1 ID-Based Signature (IBS)

*Definition*: The four algorithms that define the IBS scheme are *Setup, Key Extraction, Signature Generation*, and *Signature Verification*.

The *Setup* and *Key Extraction* processes are executed before the deployment of the sensor network. Usually, the sink node or base station takes the role of a PKG and performs the initialization process through *Setup* and key generation through the *Key Extraction* process.

*Setup*: Given a security parameter $x \in Z_q^*$ to this algorithm, it outputs system parameters $PP$ and a master secret $M_s$. The master secret is only known to PKG.

*Key Extraction*: Given a user's identity $ID_i$ and the master secret $M_s$ to this algorithm, it outputs the private-key $d_{ID_i}$.

$ID_i$, $d_{ID_i}$, and $PP$ are prestored in a sensor node before deployment.

*Signature Generation*: Given the message $m \in M$ and the private key $d_{ID_i}$, this algorithm outputs the signature $\sigma$.

*Signature Verification*: Given a message $m \in M$, the signer's identity $ID_i$, signature $\sigma$, and system parameters $PP$, this algorithm returns valid (1) or invalid (0).

### 5.3.2 ID-Based Online/Offline Signature (IBOOS)

The IBOOS algorithm is usually used for authenticated broadcast.

*Definition*: The following five algorithms define the ID-based online/offline signature (IBOOS) scheme.

*Setup* and *Key Extraction* are the same as defined in IBS.

The signature generation process is divided into two processes, namely, *Offline Signature Generation* and *Online Signature Generation*.

*Offline Signature Generation*: Given the system parameters $PP$ and the signing key $d_{ID_i}$, execution of this algorithm results in a partial offline signature $\sigma_{Off}$. This phase is usually executed on a resource abundant device such as a sink node or a base station before the message to broadcast becomes available. In deployment scenarios where the sensor network is divided into clusters, cluster heads could take the responsibility of executing this algorithm. The resulting signature is stored on each sensor node before it is deployed.

*Online Signature Generation:* This algorithm is executed whenever a sensor node must quickly report an event to its respective sink node or its cluster head. It is performed on resource constraint devices such as sensor nodes. Given $m \in M$, $\sigma_{Off}$, and the time stamp $T_i$, the algorithm generates an online signature $\sigma_{On}$. The reuse of the partial offline signature $\sigma_{Off}$ computed in the offline phase will considerably reduce the energy consumption in generating $\sigma_{On}$. Hence, the nodes only need to carry out a small number of energy-efficient computations. The final broadcast message includes $m \in M$, $T_s$, $ID_i$, and signature $\sigma_{On}$.

*Online Signature Verification*: When a sensor node receives a broadcast message, its first task is to check the time stamp to avoid any kind of a replay attack. Upon verifying the freshness of the time stamp, it runs the *Online Signature Verification* algorithm. This algorithm takes $m \in M$, $ID_i$, $\sigma_{On}$, and $PP$, and returns valid (1) or invalid (0). If required, the node could rebroadcast the message with its signature and a new time stamp.

### 5.3.3 BNN-IBS Scheme

Bellare et al. (2004) proposed the first ECC-based IBS scheme called BNN-IBS, with provable security. The implementation is as follows:

*Preliminaries*: Let $E$ be an elliptic curve defined over a prime finite field $F_q$ and denoted as $E/F_q$. Let $E(F_q)$ denote the group of points formed by $E/F_q$, which includes the point at infinity $\mathcal{O}$. Hence, $E(F_q) = \{(x,y): x,y \in F_q; (x,y) \in E/F_q\} \cup \{\mathcal{O}\}$.

*Setup*: Given the security parameter $x \in Z_p$, the following steps are carried out by the PKG.

- Let $H_1 : \{0,1\} \times G_1^* \rightarrow Z_p$ and $H_2 : \{0,1\}^* \rightarrow Z_p$.
- Specify $E/F_q$ the elliptic curve $E$, where $P \in E(F_q)$ is a point of order $p$ and $G$ is a group generated by $P$. Let the system public key $P_0 = xP \in PP$.
- The public parameters include $PP = \langle E / F_q, P, p, P_0, H_1, H_2 \rangle$.

*Key Extraction*: Given an end-user's identity (ordinary node's identity) *ID*, the PKG (sink node) validates the identity and generates the private key based on the Schnorr signature scheme (Schnorr 1991):

- Choose at random $r \in Z_p$ and compute $R = rP$.
- Use system secret key $x$ to compute $s = r + cx$, where $c = H_1 (ID\|R)$.

The pair $(R,s)$ is sent to the end user via a secure channel.

*Signature Generation*: The signer generates the signature $\sigma$ as follows:

- Choose at random $y \in Z_p$ and compute $Y = yP$.
- Compute $z = y + hs$, where $h = H_2 (ID, m, R, Y)$.

Then, the sender's signature on message $m$ is the tuple $\langle R,Y,z \rangle$. Hence, $m$ and the tuple are sent to the receiver.

*Signature Verification*: The verifier verifies the signature as follows:

- Compute $h = H_2\ (ID,\ m,\ R,\ Y)$ and $c = H_1\ (ID\|R)$.
- Check whether the equality

$$
\begin{aligned}
zP \overset{?}{=} Y &+ h(R + cP_0) \\
&= yP + h(rP + cxP) \\
&= P(y + h(r + cx)) \\
&= P(y + h(s)) \\
&= zP
\end{aligned}
\tag{5.1}
$$

holds true or not to determine the validity of the signature.

Although BNN-IBS is a non–pairing-based approach, it is still not efficient in terms of the signature size because it includes two points $R$ and $Y$ and an integer element. Hence, it would take at least two IEEE 802.15.4 packets to carry this signature scheme. A variant of BNN-IBS—$v$BNN-IBS—has been proposed by Cao et al. (2008), which reduces the signature size when compared to BNN-IBS. $v$BNN-IBS is reviewed in Section 5.3.3.1.

## 5.3.3.1 vBNN-IBS

The *Setup* and *Key-Extraction* processes are similar to the processes mentioned in BNN-IBS.

*Signature Generation*: The signer generates the signature $\sigma_s$ as follows:

- Choose at random $y \in Z_p$ and compute $Y = yP$.
- Compute $z = y + hs$, where $h = H_2\ (ID,\ m,\ R,\ Y)$.

Then, the sender's signature on message $m$ is the tuple $\langle R, h, z \rangle$. Hence, $m$ and the tuple are sent to the receiver.

*Signature Verification*: Given $\langle R, h, z \rangle$, $ID$, and the message $m$, the receiver computes $c = H_1\ (ID\|R)$ and verifies the signature as shown below:

$$
h \overset{?}{=} H_2(ID, m, R, zP - h(R + cP_0))
\tag{5.2}
$$

In comparison to BNN-IBS, $v$BNN-IBS generates a signature with a smaller size when compared to BNN-IBS while achieving the same computational complexity. As a result, this scheme is easily applicable in multiuser broadcast authentication in WSN.

## 5.3.4 Description of Identity-Based Multiuser Broadcast Authentication in Wireless Sensor Networks (IMBAS)

In this scheme, sensor node-to-node communication uses algorithms mentioned in $v$BNN-IBS algorithms (Section 5.3.3.1). One of the crucial functions of the sink node is to broadcast data related to routing-tree construction that would result in a larger message size. Hence, to improve efficiency it is recommended to use the Schnorr signature scheme with partial message recovery to secure sink node to ordinary node communication.

### 5.3.4.1 IMBAS for Node-to-Node Communication

IMBAS consists of *System Initialization, User Addition, Message Broadcast and Authentication*, and *User Revocation*

*System Initialization*:

The sink node selects $x \in Z_p$ and the system public key $P_0 = xP$. System parameters $PP = \langle E / F_q, P, p, P_0, H_1, H_2 \rangle$ are loaded into each sensor node of the network.

*User Addition*:

The sensor nodes in the network choose their unique IDs and send a request for a private key corresponding to this ID. The sink node verifies the identity and then generates a private key $(R,s)$ for the respective sensor node. The key extraction algorithm used in this scheme is similar to the $v$BNN-IBS scheme.

*Authentication*:

A sensor node in the network would broadcast a message that contains the following information: <$m,tt,ID$,Sig{$m,tt,ID$}>, where the Sig indicates the $v$BNN-IBS signature algorithm. In addition, the $tt$ refers to the time when the signature was generated.

*Verification*:

Upon receiving the authenticated broadcast message, the respective end users validate the time stamp $tt$ and then verify the broadcast signature using the verification algorithm, as mentioned in Section 5.3.3.1.

### 5.3.4.2 IMBAS for Sink-to-Node Communication

Because the sink uses the Schnorr signature to generate a private key pair for each user and the Schnorr signature is more efficient than $v$BNN-IBS, it will still use the Schnorr signature to secure its broadcast. The messages broadcasted by the sink are inclusive of certain data value, such as the data distributed for routing-tree construction. Hence, the message size will be longer. To reduce the overall message size, the Schnorr signature is combined with the message recovery technique proposed in Naccache and Stern (2000) and Nyberg and Rueppel (1993). Therefore, to broadcast a message $M$, the sink uses the following algorithm to generate the signature:

*Signature*:

*Step 1*: Prepare the broadcast message $\langle ID_{sink}, tt, M \rangle$ and break it into two parts, $M_1$ and $M_2$, where $|M_1| \leq 10$ bytes and $M_2$ is inclusive of $ID_{sink}$ and $tt$.

*Step 2*: Choose at random $y \in Z_p$ and compute $Y = yP$.

*Step 3*: Encode-and-hash $Y$ into an integer $i$ to $H_1$.

*Step 4*: Add proper redundancy to $M_1$ according to certain standards, such as the *IEEE Standard-P1363a: Standard Specifications for Public-Key Cryptography*, and the resulting value is $f_1$.

*Step 5*: Compute $f_2 = H_1(M_2)$.

*Step 6*: Then compute $c = i + f_1 + f_2$ mod $p$, such that $c \neq 0$. Otherwise start from Step 1.

*Step 7*: Compute $d = y - cx$ mod $p$, and output $(c,d)$ as the signature.

*Step 8*: Then the sink broadcasts $\langle M_2, c, d \rangle$.

*Verification*:

Upon receiving, $\langle M_2, c, d \rangle$, a sensor node checks if $tt$ in $M_2$ is fresh to avoid any kind of replay attack. The following steps are taken for verifying the signature:

*Step 1*: Discard the message if $c \notin [1, p-1]$ or $d \notin [1, p-1]$.

*Step 2*: Compute $Q = dP + cP_0$.

*Step 3*: Discard the message if $Q = \mathcal{O}$.

*Step 4*: Encode-and-hash $Q$ into an integer $i$.

*Step 5*: Compute $f_2 = H_1(M_2)$, and compute $f_1 = c - i - f_2$ mod $p$.

*Step 6*: Discard the message if the redundancy of $f_1$ is incorrect.

*Step 7*: Otherwise, accept the signature and reconstruct $\langle ID_{sink}, tt, M \rangle = M_1 \| M_2$.

## 5.3.4.3 User Revocation

User revocation is achieved by simply broadcasting the identity of the revoked users by the sink node. If the identity already exists in the revocation list, the broadcast message is ignored. To overcome a physical node capture attack, IMBAS recommends a password-based protection approach to safeguard the private keys. The node first chooses a password $PW$, and then computes $R' = H_1(PW)^{-1}R$ and $s' = H_1(PW)^{-1}s$. $(R',s')$ is stored in the user's physical device instead of $(R,s)$. If the node wants to use the private key pair, it should first key in PW. $(R,s)$ will be recovered from the stored $(R',s')$ only when the correct PW is provided. Hence, retrieving $(R,S)$ from $(R',s')$ is equivalent to solving the Elliptic Curve Discrete Log Problem (Chapter 3).

### 5.3.4.4 Security Analysis for IMBAS

If an adversary A can forge a valid $v$BNN-IBS tuple $\langle R,h,z \rangle$ for message $m$ and identity $ID$, then A can easily compute $Y$ as shown below:

$$Y = zP - h(R + H_1(ID \| R)P_0) \tag{5.3}$$

Then, A outputs the forged signature $\sigma' = \langle R,Y,z \rangle$ along with $m$ and $ID$. Upon receiving the signature $\sigma'$ along with $m$ and $ID$, he computes $h = H_2(ID,m,R,Y)$ and $c = H_1(ID\|R)$ and verifies if

$$zP \overset{?}{=} Y + h(R + cP_0) \tag{5.4}$$

Substituting for $Y$ from Eq. (5.3)

$$zP \overset{?}{=} zP - h(R + cP_0) + h(R + cP_0) \tag{5.5}$$

Let $t$ be the time taken to forge a valid $v$BNN-IBS tuple $\langle R,h,z \rangle$. Then, the time taken to forge $\langle R,h,z \rangle$ for a message $m$ and identity $ID$ with certain probability is $t + 3t_m + 2t_a$, where $t_m$ is the time taken to compute the point multiplication and $t_a$ is the time taken to compute the point addition. Hence, the conclusion is that $v$BNN-IBS is existential unforgeable if BNN-IBS is existential unforgeable. In addition, IMBAS uses Schnorr's signature with partial message recovery, which has been proven secure (Nyberg and Rueppel 1993; Naccache and Stern 2000). As a result, A is unable to forge packets or modify broadcasted messages.

DoS attacks on WSN are catastrophic because they lead to resource exhaustion. IMBAS prevents DoS attacks by dropping packets whose signature fail to verify. In addition, this scheme enforces a limit on the number of verifications. Consequently, if the number of checks exceed the threshold limit, the verifying node can inform the sink node about the adversary for further investigation and revocation of the malicious node A. Concerning scalability, IMBAS can easily incorporate new nodes by preloading them with system parameters.

Although IMBAS proposes an authentication scheme for WSN, it does not address authenticated session key establishment between the user and the sensor node. The next section describes a pairing-free ID-based one-pass authenticated key establishment between a user and a node.

## 5.3.5 ID-Based One-Pass Authenticated Key-Establishment Protocol for WSN*

### Rehana Yasmin, Eike Ritter, and Guilin Wang

Authenticated session key-establishment techniques have been an integral part of any legacy network such as IP networks. These protocols provide the concerned communicating parties with an authentic shared session key in a secure way. For WSN, the use of such session keys is particularly important because there is a need to securely exchange data between the sink and the leaf nodes. In addition, external users also access the aggregated data from the sink or from the leaf nodes directly. However, the use of a two-pass key-establishment protocol (e.g., the Diffie-Hellman key exchange mechanism) in a resource constraint environment such as WSN could incur increased storage, communication, and computational costs. Hence, there is a need for an authenticated session key-establishment scheme that should provide adequate security and be resource efficient in an environment such as WSN.

An easy alternative that could be resource efficient and provide reasonable security is a one-pass authenticated key-establishment method. As the name states, the sender generates an ephemeral private key, and its corresponding public key is only sent once to the receiver. Subsequently, both parties compute the shared session key using the ephemeral key and their private key.

ID-based one-pass authenticated key-establishment protocol by utilizing the four phases: *System Initialization*, *Private Key Generation*, *User Registration*, and *Key Establishment*. The first two phases are performed once, before the deployment of the sensor network. In an ID-based cryptosystem, a PKG computes the private keys corresponding to the IDs. In WSNs, the base station (a resourceful device) is considered as trustworthy. In this scheme, the base station plays the role of PKG and computes the private keys for sensor nodes and users.

### 5.3.5.1 System Initialization

In this phase, the Setup algorithm runs on the sink node and generates the system parameters, including the master public key (*Puk*) and the corresponding master secret key (*Prk*), by using a security parameter $k$.

This algorithm performs the following steps:

- Specify $q$, p, $E/F_p$, $P$ and $G$, where $q$ is a large prime number and $p$ is the field size, $E/F_p$ is an elliptic curve $E$ over a finite field $F_p$, $P$ is a base point of order $q$ on the curve $E$, and $G$ is a cyclic group of order $q$ under the point addition "+" generated by $P$.
- For *Prk*, $s \in Z_q^*$, compute *Puk* as $P_{PKG} = sP$.
- Choose one hash function $H : \{0,1\}^* \times G \rightarrow Z_q^*$.
- Choose one key derivation function $\chi : G \rightarrow \{0,1\}^k$
- Output system parameters $\{q, p, E / F_p, P, G, P_{PKG}, H, \chi\}$ and keep $s$ secret.

* Originally published as Rehana Yasmin, Eike Ritter, and Guilin Wang, "Pairing-Free ID-based One-Pass Authenticated Key Establishment Protocol for Wireless Sensor Networks" presented at the *Fifth International Conference on Sensor Technologies and Applications* from August 21–27, 2011.

### 5.3.5.2 Private Key Generation

In this phase, the *Extract* algorithm runs on the base station (before deployment) and computes the private keys of all sensor nodes corresponding to their *IDs*. This algorithm takes *Prk* and a sensor node's *ID* as input and generates a private key corresponding to that ID, using the well-known Schnorr signature. For a sensor node *I* with identity $ID_i$, this algorithm performs the following steps:

- For $r_i \in Z_q^*$, compute $R_i = r_i P$ and $c_i = H(ID_i, R_i)$.
- Compute private key as $s_i = c_i s + r_i$.
- Output $(s_i, R_i)$, where $s_i$ is secret while $R_i$ is public. Here, the private key $s_i$ is the Schnorr signature on the *ID* of the node signed with the private key of the PKG. *IDs* corresponding to private keys and system parameters are stored on sensor nodes before deployment. Hence, every sensor node *i* stores $\{ID_i, s_i, R_i\}$ and system parameters.

### 5.3.5.3 User Registration

This phase is repeated every time a new user is registered with the system. In this phase, the *Extract* algorithm runs on the base station and computes the private key for a user *U* corresponding to his identity $ID_u$ in the same way as it is computed for sensor nodes in the *Private Key Generation* phase. The base station, which runs this algorithm, sends the private key to the user via a secure channel. Hence, every user *U* gets $\{ID_u, s_u, R_u\}$ and system parameters.

### 5.3.5.4 One-Pass Authenticated Session Key Establishment

Whenever a user wants to access data from sensor nodes, he establishes a session key with the sensor node in his range, after successfully authenticating himself to the sensor node. Whether the user query is processed by a single sensor node or a set of sensor nodes is related to the topic of query processing in wireless sensor networks and is not addressed herein. We now describe an ID-based one-pass session key-establishment protocol between a user *U* and a sensor node *I*. Figure 5.3 describes the steps of the Authenticated One-Pass Session Key-Establishment Protocol with Key Confirmation protocol.

The user *U* chooses at random $t \in Z_q^*$ as ephemeral key and computes $y = t s_u$ and $L = yP$. *U* signs the ephemeral public key *L* together with $ID_u$, $ID_i$ and $T_S$ and sends $[L, ID_u, ID_i, T_S, Sig_{su} (L, ID_u, ID_i, T_S)]$ to the sensor node *I* in his range. Here $T_S$ is the current time stamp to avoid a replay attack and $Sig_{su} (L, ID_u, ID_i, T_S)$ is a signature signed by *U* using his private key $s_u$. Computing *y* from *L* is the so-called *Elliptic Curve Discrete Logarithm* (ECDL) problem, which is intractable.

The sensor node *I* first checks the time stamp $T_S$ to avoid the verification of a replayed message. If this is a fresh message, *I* verifies the signature $Sig_{su} (L, ID_u, ID_i, T_S)$. Successful signature verification implies that the message is actually sent by the

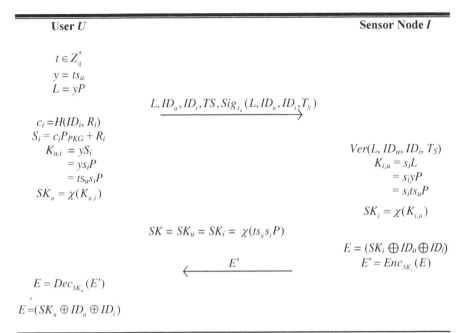

**Figure 5.3.** Authenticated One-Pass Session Key-Establishment Protocol with Key Confirmation

user $U$ and is fresh. Hence, $I$ accepts the message; otherwise, the protocol is terminated at this stage. Next, the sensor node $I$ computes the shared secret $K_{i,u}$ as $K_{i,u} = s_i L (=s_i t s_u P)$ and deletes $L$.

The user $U$ computes the same shared secret $K_{u,i}$ as $S_i = c_i P_{PKG} + R_i$, where $c_i = H(ID_i;R_i)$ and $K_{u,i} = y S_i$ $(=ts_u s_i P)$. $U$ then deletes $L$, $t$, and $y$. Both parties then compute the shared session key as $SK = \chi(K_{u,i}) = \chi(K_{i,u}) = \chi(ts_u s_i P)$, where $\chi$ is the key derivation function. However, there is no guarantee that at the end of the secure run of the protocol both parties compute the key. Indeed, in any key-establishment protocol, the sender of the last message cannot confirm that the last message is received by the destined receiver. The user may successfully finish the protocol with a key output. Although the adversary is not able to learn the computed key, the sensor node might not receive the user's message and, consequently, might not be able to compute the key. The assurance against this scenario is achieved via an authenticated key-establishment protocol with key confirmation (AKC). This is usually achieved by adding a key confirmation message to the authenticated key-establishment protocol after the key has been established.

Hence, after both parties establish the session key, the key-establishment algorithm proceeds as follows:

After key computation, the sensor node $I$ performs the following steps:

- Computes the *XOR* of its computed key $SK_i$ with $ID_u$ and $ID_i$, as follows: $E = (SK_i \oplus ID_u \oplus ID_i)$.

- Encrypts $E$ with $SK_i$ using a secure symmetric encryption algorithm, i.e., $E' = Enc_{SK_i}(E)$ and sends $E'$ to $U$.
- After $U$ receives $E'$, he performs the following steps:
    i. Decrypts $E'$ using his computed key $SK_u$ to obtain $E$, i.e. $E = Dec_{SK_u}(E')$
    ii. Checks whether $E \overset{?}{=} (SK_u \oplus ID_u \oplus ID_i)$

Successful verification implies that both parties have computed the shared session key. As the user does not expect to receive a message from the sensor node to compute the key, he does not need to send a key-confirmation message to the sensor node.

### 5.3.5.4.1 Authentication, Key Establishment, and Query Privacy

To retrieve data from the sensor network, each end user authenticates himself to a nearby node. He then establishes a session key and sends the query to it. The sensor node, after successful user authentication and session key establishment, processes the received user query, encrypts the query results, and sends them back to the user. For privacy reasons, the user query needs to be encrypted in some situations because users may not be willing to disclose their areas of interests (Carbunar et al. 2007). Due to the one-pass key establishment, query privacy can also be provided by the proposed protocol, as follows: The user computes his shared session key, encrypts his query using computed session key, and sends his signed ephemeral public key to the sensor node, together with his encrypted query, in a single message.

The sensor node first authenticates the user by verifying the signature. If the signature verification fails, the protocol terminates here. Otherwise, the sensor node computes the same shared session key, decrypts the user query, processes it, and sends the encrypted query results back to the user. Thus, only a single message is exchanged for authentication, key establishment, and encrypted query transmission achieving transmission efficiency.

### 5.3.5.4.2 ID-Based Signature Scheme

To sign the ephemeral public key, any secure IBS scheme with the same ID-based parameters can be used (see Section 5.3.1). In addition, a variant of BNN-IBS—$v$BNN-IBS (see Section 5.3.3.1)—can be used to provide broadcast authentication.

### 5.3.5.4.3 Distributing the Public Information $ID_i$ and $R_i$

One possible question might be how a user can obtain $ID_i$ and $R_i$, the public information of a sensor node $I$. As the user is equipped with a resourceful device, it can store the $ID_i$ and $R_i$ pairs of the sensor nodes in the user's range. In 160-bit ECC settings, the size of the $ID_i$ and $R_i$ pair is about 25 bytes. For approximately 5000 sensor nodes in the user's range, the total storage required is about 125 KB. This is an acceptable storage overhead on a resourceful user device to provide security with efficiency on resource-constrained sensor nodes. The user can also obtain a $ID_i$ and $R_i$ pair from the base station through any other means—e.g., the Internet—before making a query to $I$. Note that here $ID_i$ and $R_i$ are two identity elements of $I$ and not the public keys, which is different from the traditional public key crypto system in which public keys are verified using the signed certificates. Here, if someone tries to use a fake $ID_i$ and $R_i$ pair, he cannot generate a corresponding private key $s_i$, which is generated using $Prk$. Generating such a valid triplet without $Prk$ would be equivalent to forging the Schnorr signature.

This setup allows the construction of efficient pairing-free ID-based schemes and handles the problem of public keys/certificates.

### 5.3.5.5 Security Analysis

The security of this protocol is formally analyzed using the reductionist proof technique under the standard Computational Diffie-Hellman (CDH) assumption. By assuming that the CDH assumption holds in $G$, we show that the proposed protocol is secure in the ID-eCk model (Gorantla et al. 2008). Due to space limitations, the security model and rigorous proof are omitted in this chapter.

Here, we informally discuss five security attributes pertaining to the proposed protocol.

*Authentication.* The proposed protocol provides the required authentication. There is only one message exchanged, and that is sent by the user. Authentication of that single message is achieved by the verification of signature signed by the user. It is not feasible for an adversary to sign a message on behalf of a user without knowing the user's private key. Successful signature verification by the sensor node $I$ proves that the ephemeral public key is actually sent by a legitimate user $U$. On the other side, $S_i(= s_iP)$ computed from $I$'s public information assures the user that the session key is, in fact, established with $I$. Only the sensor node $I$ with the valid corresponding private key $s_i$ can compute the same session key. Authentication avoids the chances of the adversary mounting a man-in-the-middle attack.

*Key confidentiality.* After the successful key establishment between a sensor node and a user, the public parameters and the ephemeral public key $L(= ts_uP)$ are the only information available to the adversary. However, the adversary cannot compute the user $U$'s private key $s_u$ and/or ephemeral private key $t$ from $L$ since we assume there is no polynomial time algorithm to solve the ECDL problem. Furthermore, he cannot compute the shared secret $ts_us_iP$ because it requires the knowledge of private keys of both the sensor node and the user. Hence, the key is computable only by the user $U$ and the sensor node $I$.

*Key compromise.* The random value for the ephemeral private key $t$ is separately generated for each session. Therefore, the established session key is computationally different for different sessions. A session key established between a compromised sensor node and a user would not enable an adversary to compute or learn any other session key established between any other legitimate sensor node and a user. Furthermore, it would not enable an adversary to learn the user's private key $s_u$ from $L$ due to the intractability of the ECDL problem. In fact, the proposed protocol guarantees that the communication between an uncompromised sensor node and a user cannot be exposed, irrespective of the number of other nodes that are compromised.

*Key confirmation.* In the proposed protocol, the key confirmation message $E_0$ provides the explicit key confirmation.

The sensor node computes $E'$ and sends it to the user so that the user can be assured that the sensor node has received the user's ephemeral public key and successfully computed the session key. However, the user does not expect to receive any message from the sensor node for key establishment because he can compute the same session key by himself. Hence, the user does not need to send a key confirmation message back to the sensor node.

*Replay attack.* In a replay attack, an adversary replays the previous successful user request to either establish a session key with the sensor node or to waste sensor node resources by the request verification. In the proposed protocol, because of the user's signed message, the adversary will not be able to authenticate successfully and establish a key. Furthermore, the time stamp $T_S$ provides freshness. The sensor node checks the time stamp before the signature verification to avoid the verification of a replayed request message. Depending on the transmission delay imposed by the communication channel between the user and the sensor node, the sensor node sets a time threshold, leaving a potential attacker little time to mount a replay attack.

## 5.3.6 TinyIBE Scheme

In this model, the IBE scheme is implemented on a hierarchical architecture discussed in Section 1.3.1.2. Let $H$ be the group of static sensor nodes that takes the role of a cluster head and $L$ be a group of static ordinary sensor nodes. The sink node is responsible for issuing unique identities for each sensor node. As the $H$-sensors assume the role of a cluster head, the model assumes that they have greater computational and storage capacity when compared to $L$-sensor nodes and, hence, can store the entire $L$-sensor node *IDs* belonging to their respective clusters. In addition, maintaining a group of node *IDs* is much more efficient than storing their public key certificates.

*Setup*: Let $E$ be an elliptic curve defined over a finite field $F_q$ and $s \in Z_q^*$ be a master secret. Let $P$ and $Q \in E(F_q)$ and have order $r$, where $Q = sP$. Let $H_1 : \{0,1\}^* \to Z_q^*$ and $H_2 : F_q \to \{0,1\}^n$ for some $n$. Let all the system parameters $PP = \langle E / F_q, Q, P, g, n, H_1, H_2 \rangle$ be preloaded in a sensor node, where $g = e(P,P)^w$.

*Extract*: Let the sink node issue unique identities $ID_i$ for each node in the network. Let the identity of each $H$-sensor node be mapped to a hash value such that $a_i = H_1(ID_{H_i})$, and let $D$ be the corresponding private key such that $d_i = \left(\dfrac{1}{s + a_i}\right) P$.

*Encrypt*: Every *L*-sensor selects random integers $w \in Z_q^*$ and $t \in Z_q^*$, where *t* is the session key. The *L*-sensors then encrypt the session key as shown below:

$$C_1 = w(Q + a_iP) \text{ and } C_2 = t \oplus H_2(g^w) \tag{5.6}$$

The cipher texts $C_1$ and $C_2$ are then sent to their respective *H*-sensor nodes.

*Decrypt*: On receiving the encrypted message, each *L*-sensor decrypts the session key as shown below:

$$t = H_2(e(d_i, C_1)) \oplus C_2 \tag{5.7}$$

Using the bilinear property, the equation

$$e(d_i, C_1) = e\left[P, (Q + a_iP)^{\frac{w}{s+a_i}}\right] = e(P, P)^w = g^w$$

In this model, broadcast authentication is achieved using the Elliptic Curve Digital Signature Algorithm (ECDSA). Because only the *H*-sensor nodes are in possession of their private keys $(d_i)$, the retrieval of session key $(t)$ by these nodes provides a means of implicit authentication between *L*-sensors and their respective *H*-sensors. If an *H*-sensor node receives a request from an *L*-sensor node whose node ID is not preloaded, the *H*-sensor would contact the sink node for further verification of the *L*-sensor's node ID. In addition, periodic update of node IDs between the sink node and *H*-sensors is carried out to prevent intruders.

## 5.4 Revocation

Revocation is discussed in detail in Chapter 6 (Identity-Based Key Distribution in Wireless Sensor Networks). To revoke a compromised sensor node in WSN, the sink node or cluster head will simply send an authenticated broadcast message. This revoked message will contain the identity of the compromised node. Hereafter, if any node in the network receives a message signed by the compromised node, it would just reject the message and not waste its energy in verifying the message's authenticity. In this case, it is assumed that the adversary is capable of compromising a few sensor nodes in the network. However, if most of the nodes are compromised, it could lead to a complete downfall of the security system.

## 5.5 User Authentication

To interact with sensor nodes directly, end users should first register with the sink node or base station and obtain their respective private keys and system parameters.

Subsequently, if an end user is required to interact with the sensor nodes, he will need to send an ID-based digital signature along with the requested information. Upon successfully verifying the end user's signature, the node and the user would then establish the session key to be used for future encrypted communication.

The system initialization and key-generation processes are similar to the schemes described in IBS and IBOOS schemes.

*User registration*: This procedure is executed whenever a new user wishes to interact with the deployed sensor network. The user submits his identity and the sink node or the base station authenticates his identity based on some previously shared information. (The details of the procedure used to authenticate the end user are beyond the scope of this book.) After authenticating the user, the sink node computes his private key and sends the key along with system parameters.

*User authentication and verification*: To interact with the nearby deployed sensor nodes, the end user sends a signed request and his identity, along with a time stamp. Upon receiving the request, the nodes verify the freshness of the time stamp and the authenticity of the message. If the verification process is successful, the nodes then calculate the session key.

*Session key establishment*: Session key establishment is discussed in detail in Chapter 6. One of the important requirements for establishing a session key between the end user and the resource-constraint sensor nodes is computing the key in an energy-efficient way. Among the several key-establishment algorithms, identity-based one-pass key-establishment protocol could be a suitable choice in such networks (Gorantla et al. 2008). In this protocol, the number of messages exchanged is considerably low because only one party computes the ephemeral key and sends it to the other party. This key could be sent along with the initial authenticated message with a time stamp, which may further reduce the overhead in communication.

*User revocation*: In end-user revocation, two cases exist, as follows:

- To revoke a malicious user
- To revoke a user whose access time period has expired

In the first case, the sink node simply broadcasts an authenticated message with the identities of end users who should be revoked. It also includes the expiration time of the identities that have been blacklisted. Consequently, if an end user sends an authenticated message requesting access to the aggregated information by the sensor node, the node checks to see if his identity is blacklisted. If the identity appears in this list, the request will be dropped by the sensor. In addition, the procedure for storing identities on nodes until the private key expires will not impose an unreasonable amount of overhead on the sensor nodes. To increase the efficiency of storing revoked identities, the expiration time should be short enough so that the nodes do not store the blacklisted identities for a longer period.

In the second case, while calculating the private key for the end user, the sink node or the cluster head takes the expiration time as one of the input parameters. Hence, the user will not be able to interact with the sensor network after the expiration time.

## 5.6 Summary

Authentication has been one of the fundamental building blocks in security. In this chapter, we review the existing non-ID–based authentication schemes and describe their drawbacks. We then look into ID-based authentication schemes. Mainly, we discuss IBS, BNN-IBS, and its variant $v$BNN-IBS. Later, we review authenticated session key-establishment schemes in WSN and discuss a pairing-free one-pass session key-establishment scheme in WSN. In addition, we review a pairing-based encryption scheme and end the discussion with revocation schemes for end users.

## 5.7 References

Bellare, M., C. Namprempre, and G. Neven. Security proofs for identitybased identification and signature schemes. *Proceedings of Eurocrypt.* Springer-Verlag, 2004, 268–286.

Cao, X., W. Kou, L. Dang, and B. Zhao. IMBAS: Identity-based multi-user broadcast authentication in wireless sensor networks. *Computer Communication* (Elsevier) 31 (2008): 659–667.

Carbunar, B., Y. Yu, L. Shi, M. Pearce, and V. Vasudevan. Query privacy in wireless sensor networks. *Sensor, Mesh and Ad Hoc Communications and Networks.* IEEE, 2007, 203–212.

Gorantla, M. C., C. Boyd, and J. M. González Nieto. ID-based one pass authenticated key establishment. *Proceedings of the Sixth Australasian Conference on Information Security.* 2008, 39–46.

Huang, Q., J. Cukier, H. Kobayashi, B. Liu, and J. Zhang. Fast Authenticated Key Establishment Protocols for Organizing Sensor Networks. *Workshop on Sensor Networks and Applications (WSNA).* ACM, 2003, 141–150.

Karlof, C., N. Sastry, and D. Wagner. TinySec: A link layer security architecture for wireless sensor networks. *Second ACM Conference on Embedded Networked Sensor Systems.* ACM, 2004, 162–175.

Liu, D, and P. Ning. Efficient distribution of key chain commitments for broadcast authentication in distributed sensor networks. *10th Annual Network and Distributed System Security.* 2003, 263–276.

Merkle, R. C. Protocols for Public Key Cryptosystems. *Symposium on Security and Privacy.* IEEE, 1980, 122–134.

Mitzenmacher, M. Compressed bloom filters. Edited by ACM. *Transactions on Networking* 10, no. 5 (2002).

Naccache, D., and J. Stern. Signing on a postcard. *Proceedings of Financial Cryptography.* Springer-Verlag, 2000, 121–135.

Nyberg, K., and R. A. Rueppel. A new signature scheme based on the DSA giving message recovery. *1st ACM Conference on Computer and Communications Security.* ACM, 1993, 58–61.

Perrig, A., R. Canetti, J. D. Tygar, and D. Song. Efficient authentication and signing of multicast streams over lossy channels. *Symposium on Security and Privacy.* IEEE, 2000, 56–73.

Perrig, A., R. Szewczyk, V. Wen, D. Culler, and D. Tygar. SPINS: Security protocols for sensor networks. Edited by ACM. *Proceedings of MobiCom.* Rome, 2001, 189–199.

Ren, K., W. Lou, and Y. Zhang. Multi-user broadcast authentication in wireless sensor networks. *Proceedings of Sensor, Mesh and Ad Hoc Communications and Networks.* IEEE, 2012, 223–232.

Schnorr, C. P. Efficient signature generation for smart card. *Journal of Crytology* 4, no. 3 (1991): 161–174.

Tian, X., D. Wong, and R. Zhu. Analysis and improvement of an authenticated key exchange protocol for sensor networks. *Communications Letters* (IEEE) 9, no. 11 (2005): 970–972.

Wander, A. S., N. Gura, H. Eberle, V. Gupta, and S. C. Shantz. Energy analysis of public-key cryptography for wireless sensor networks. *Third IEEE International Conference on Pervasive Computing and Communications.* IEEE, 2005, 324–328.

Watro, R., D. Kong, S. Cuti, C. Gardiner, C. Lynn, and P. Kruus. TinyPK: Securing sensor networks with public key technology. *2nd ACM Workshop on Security of Ad Hoc and Sensor Networks.* ACM, 2004, 59–64.

Yasmin, R. An authentication framework for wireless sensor networks using identity-based signatures: implementation and evaluation. Special section on Trust, Security and Privacy in Computing and Communication Systems in *IEICE Transactions on Information and Systems (IEICE-2011).* Vol. E95-D, No. 01, pp. 126–133, January 1, 2012.

Yasmin, R., E. Ritter, and G. Wang. An authentication framework for wireless sensor networks using identity-based signatures. *10th IEEE International Conference on Computer and Information Technology.* 2010, 882–889.

Yasmin, R., E. Ritter, and G. Wang. Pairing-free ID-based one-pass authenticated key establishment protocol for wireless sensor networks. Presented at the *Fifth International Conference on Sensor Technologies and Applications.* August 21–27, 2011.

# Chapter 6

# Identity-Based Key Distribution in Wireless Sensor Networks

## In This Chapter

- Introduction to key distribution in WSN
- Symmetric key algorithms for WSN
- Public key-distribution schemes
- ID-based bilinear key-distribution schemes
- Matsumoto-Imai's key-predistribution scheme

## 6.1 Introduction

Recent advances in Integrated Circuit (IC) fabrication have led to the proliferation of wireless sensor networks (WSN), which comprise low-cost sensors with limited storage and processing power. WSN have applicability in diverse fields, such as: military, ocean, and wildlife monitoring; earthquake monitoring; safety monitoring in buildings; and in new smart home technology proposed by 4G technologies. However, such networks deviate from legacy-embedded wireless networks in terms of scalability, dynamic nature with regard to the addition or deletion of nodes, and deployment areas. Hence, there is a greater challenge in providing security, by taking such harsh operational requirements into consideration. One such challenge is in key distribution and its management. In addition, the lack of *a-priori* information about the topology of WSN makes key management fairly complex. Key distribution provides communication secrecy (confidentiality) and authentication among sensor nodes, and key revocation refers to the task of removing compromised keys from the network. Key distribution can be further divided into symmetric and asymmetric key-distribution protocols.

In recent years, considerable work has been done in proposing new symmetric key-distribution protocols in WSN, but less effort has been invested in the area of asymmetric key-distribution algorithms in WSN, which have low computational and storage requirements. Of late, significant work has been done to show the applicability of implementing binary-field algorithms on sensor nodes (Aranha et al. 2010). Consequently, such implementations have resulted in considerable reductions in computational time and memory access.

## 6.2 Background

Key-distribution techniques need to meet several requirements, including the following: support for in-network processing, capacity to self-organize and minimally incorporate authentication, confidentiality, integrity, flexibility, and scalability. Key-establishment techniques should not be judged solely on their ability to provide secrecy for transmitted messages. They should provide resistance against malicious node replication, revoke compromised nodes, and provide resilience by ensuring that secret information about other nodes is not revealed if one of the sensors in the network is compromised.

Perrig et al. (2002) proposed a link-layer security architecture that includes two link-layer protocols, namely, Secure Network Encryption Protocol (SNEP) and μTESLA. SNEP supports two-party authentication and data freshness, along with confidentiality. Although it is the second part of SPINS (Security Protocols for Network Sensors), μTESLA supports authenticated broadcast for sensor networks. A multilevel key-chain scheme for μTESLA was proposed by Liu and Ning (2003). The first probabilistic key-predistribution scheme for pairwise key establishment was proposed by Eschenauer and Gligor (2002) discussed in Section 6.3.4.1. This approach is further extended by Chan et al. (2003) who proposed the $q$-composite keys scheme (Section 6.3.5). This scheme introduces the multipath key-reinforcement scheme by establishing link keys through multipaths and provides security against a node-capture attack. Furthermore, this model introduces a random pairwise key scheme that facilitates node-to-node authentication between communicating entities (Section 6.3.6). Du et al. (2003) showed that using the prior node-deployment knowledge and a smart key-ring setup would lead to a higher link probability between any two nodes.

## 6.3 Key-Distribution Schemes

The past decade has seen a considerable amount of research on tackling the problem of key distribution in WSN. In short, the problem refers to the task of preloading the sensor nodes with the necessary cryptographic information without any prior knowledge of the deployment topology, and with the intention that the nodes are able to communicate securely with each other after deployment. Furthermore, the scheme should work with the increase in scalability (addition of new nodes to the network) and should adhere to the low computational and storage requirements.

In general, key-distribution schemes in WSN can be broadly classified into four classes: symmetric key algorithms, trusted server mechanisms, random key-predistribution schemes, and public key algorithms. In the following sections, we review a few existing key-distribution schemes in WSN.

## 6.3.1 Symmetric Key Algorithms

In this class, a single shared key is used to perform the encryption and decryption operations in a communication network.

### 6.3.1.1 Fully Pairwise-Shared Keys

In this scheme, every node in the network shares a unique, preshared, symmetric key with every other node in the network. The keys are preloaded into the senor nodes before deployment. Hence, in a network of n nodes, there would be a total of $n(n–1)/2$ unique keys. Subsequently, every node stores $n$-1 keys, one for each of the other nodes in the network. In this class of protocols, the compromise of a few sensor nodes will not result in the complete collapse of the entire network. However, the applicability of this approach in large sensor networks is not pragmatic, as each node would need to store $n$-1 keys, thus resulting in the rapid exhaustion of its limited memory space. In addition, nodes usually communicate with their immediate one-hop neighbors, thereby eliminating the need to establish unique keys with every node in the network. Although symmetric key algorithms are limited in terms of key distribution, they provide basic cryptographic primitives, which could be used in combination with asymmetric key cryptographic algorithms.

## 6.3.2 Trusted Server Mechanisms

In this category, key distribution is done via centralized trusted servers, which are usually static in nature. In WSN, the sink node or the base station can act as a key-distribution center (KDC). Usually, unique symmetric keys are shared between the sink node and the ordinary nodes. If two nodes were to communicate with each other, they would first authenticate with the base station, after which the base station generates a link key and sends it securely to both parties.

An example of a base-station-mediated key-agreement protocol is the Security Protocol for Sensor Networks: SPINS (Perrig et al. 2002). Using this protocol, only one unique single key is preloaded in every node of the network. Hence, a node capture will not result in the total compromise of the network. In addition, centralized revocation is possible through authenticated unicasts from the trusted base station. The main drawback of this scheme is that the trusted base station represents a single point of compromise for security information, and may also induce a focused communication load centered on the base station, which may lead to early battery exhaustion for the nodes closest to the base station. Another concern is that certain networks do not have a suitable, highly functional, and tamper-proof device that can be used as a secure KDC.

### 6.3.3 λ-Secure n × n Key-Establishment Schemes

Blom (1984) and Blundo et al. (1992) addressed the problem of key distribution and key establishment between all pairs of $n$ principals. Although these schemes were originally intended for group keying in traditional networks, and not for sensor networks, they are included here because of their relevance to the development of subsequent key-distribution schemes for sensor networks. The schemes of both Blom and Blundo et al. have an important resiliency property—the λ-secure property. The coalition of no more than λ-compromised sensor nodes reveals nothing about the pairwise key between any two noncompromised nodes.

The main advantage of this class of schemes is that they allow a parameterizable trade-off between security and memory overhead. Whereas the full pairwise scheme involves the storage of O($n$) keys at each node and is $n$-secure, this class of schemes allows the storage of O(λ) keys in return for a λ-secure property, and it is perfectly resilient to node compromise until λ+1 nodes have been compromised, at which point the entire network's communications are compromised.

### 6.3.4 Random Key-Predistribution Schemes

In this method, keys are predistributed by preloading random keying material on sensor nodes with the intention of establishing a common secret key between the communicating entities. Upon deployment, these nodes carry out a lookup process to see if a shared key exists between them. As keys are preloaded in a random manner, certain set of nodes may not share a common key with each other. In such cases, nodes could make use of their immediate neighbors who share keys, as bridges between the nodes that do not share a common key. One of the early, key-sharing algorithms using random graph theory was proposed by Eschenauer and Gligor (2002).

#### 6.3.4.1 Basic Random Key-Predistribution Scheme

In this scheme, let $m$ denote the number of distinct cryptographic keys that can be stored on the key ring of a sensor node. This scheme is divided into three phases as follows:

*Phase I: Key Predistribution*

In this initialization phase, a random pool (set) of keys $Q$ are picked from the total possible key space. In addition, for each node, $m$ keys are randomly selected from the key pool $Q$ and stored into the node's memory. Each of the $m$ keys have identifiers that will be used to map the keys by the receiving nodes during the discovery phase of this scheme (discussed in the next section). This set of $m$ keys is called the node's key ring. The number of keys in the key pool $|Q|$ (key pool size) is chosen such that any two random subsets of size $m$ in $Q$ will share at least one key, with some probability $p$.

*Phase II: Shared-Key Discovery*

On deployment, neighboring sensor nodes begin the discovery process to find out if they share a common key with each other; if they do, then they establish a secure link. There could be many modes for the discovery phase, such as broadcasting the list of identifiers existing in their key ring in clear text or through a challenge-response mechanism. If the probability $p$ were chosen correctly for the network's neighbor density, then the resultant graph of secure links will be connected with some high probability. The remaining links in the graph are then filled in by routing key-establishment messages along this connected network of initial secure links. From a security perspective, although this approach does not reveal any important information to the adversary, it is still susceptible to a passive traffic analysis attack.

*Phase III: Path-Key Establishment*

On completing the discovery phase, if two nodes in the network discover that they do not share a key between them, they send an encrypted message to neighbors with whom they share a key, with a request to secure a connection with the unshared node. This model assumes that after the completion of *Phase II*, there exist many keys in each key ring that can be used for third-party path-key establishment. Hence, the neighboring nodes generate pairwise keys for nodes that do not directly share a key.

Let us now find this probability $p$ that any two nodes with key ring sizes $m$ in the network share at least one common key from the pool $Q$. Let $p'$ be the probability that two nodes do not share a key between them. Then, $p$ is defined as

$$p = 1 - p' \tag{6.1}$$

In this case, keys from the key ring are drawn from $Q$ without replacement. The total number of possible key rings $t_1$ is shown below:

$$t_1 = \frac{Q!}{m!(Q-m)!} \tag{6.2}$$

Now, the total number of possible key rings that do not share a key with a particular key ring $t_2$ is the number of key rings drawn from the remaining $Q$-$m$ unused keys in the pool:

$$t_2 = \frac{(Q-m)!}{m!(Q-2m)!} \tag{6.3}$$

Then, the probability that no key is shared between any two rings is $t_2/t_1$. Hence, the probability $p$ is

$$p = 1 - \frac{t_2}{t_1} = 1 - \frac{((Q-m)!)^2}{Q!(Q-2m)!} \tag{6.4}$$

Usually, the value of $p$ is very large in comparison to $m$, and using the Sterling's approximation for $n!$, the value of $p$ is

$$p = 1 - \frac{\left(1 - \dfrac{m}{Q}\right)^{2(Q-m+0.5)}}{\left(1 - \dfrac{2m}{Q}\right)^{(Q-2m+0.5)}} \tag{6.5}$$

Figure 6.1 shows the value of $p$ for different values of $Q$ and $m$. We observe that with the increase in $Q$, there is a negligible increase in the key ring size $m$ for the same value of $p$. For example, for $p = 0.5$ and $Q = 6000$, the value of $m = 68$. Subsequently, if the pool size is increased to 10,000, for the same value of $p = 0.5$, $m$ is only increased to 95.

In this scheme, all nodes use the same key pool $Q$. This implies that the security of the network is gradually eroded as keys from $Q$ are compromised by an adversary that captures more and more nodes. In this scheme, the number of exposed keys is roughly

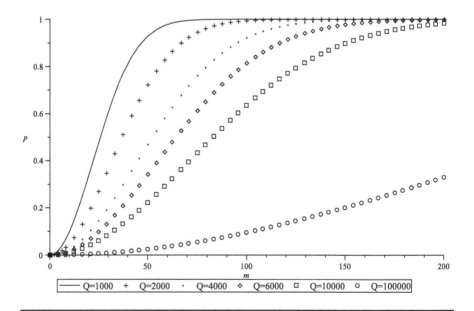

**Figure 6.1.** Probability of Sharing At Least One Shared Key Using Eschenauer and Gligor's Scheme

linear to the number of nodes compromised. This characteristic of the basic scheme motivated development of key predistribution schemes that have better resiliency to node capture. The basic scheme was extended by the $q$-composite scheme proposed by Chan et al. (2003).

### 6.3.5 q-Composite Scheme

In a $q$-composite key scheme, instead of designing for a given probability $p$ of sharing a single key, the parameters are altered such that any two nodes have a given probability $p$ of sharing at least $q$ different keys from the key pool. All $q$ keys are used in the generation of the key, which encrypts communications between sensor nodes; hence, to eavesdrop on the secured link, the adversary now has to compromise all $q$ keys, instead of just one. As $q$ increases, it is exponentially harder for the attacker to break a link by taking possession of a given set. However, increasing the probability of overlap in this fashion naturally involves reducing the size of the key pool $Q$. Thus, the smaller key-pool size makes the scheme more vulnerable to an adversary that is capable of compromising larger numbers of sensor nodes.

The key-predistribution phase of this model is similar to *Phase I* in Section 6.3.4.1, with the only exception being the key-pool size $Q$. In the shared key-discovery phase, each node must find nodes that share all common keys with each other. The discovery mechanism is similar to that of *Phase II*. Although a broadcast-based approach is susceptible to an eavesdropping attack, alternative methods that are slower but more secure are suggested where the nodes use the Merkle puzzle for key discovery (Merkle 1980). After the discovery phase, each node would be able to recognize its immediate neighboring nodes with which it would share at least $q$ keys. Subsequently, each node could establish a link between nodes that share at least $q$ keys by hashing the keys in some canonical order. For example, $K = \text{hash}(k_1 \,\|\, k_2 \,\|\, k_3 \,\|\, \dots \,\|\, k_q)$.

In this scheme, the key pool size $|Q|$ plays a critical role because with a larger $Q$, the probability of any two nodes sharing at least $q$ keys would be much less. Consequently, after bootstrapping, the network may not be connected. On the contrary, if $|Q|$ is small, the security of the network is compromised. Hence, $|Q|$ should be such that the probability of sharing at least $q$ key should be greater than or equal to the probability of successfully achieving a key setup with any of its neighbors. The approach used to calculate the probability of any two nodes sharing exactly $i$ keys $p'(i)$ is similar to calculating $p$, as shown in Eq. (6.4), and is given as

$$p'(i) = \frac{\binom{|Q|}{i}\binom{|Q|-i}{2(m-i)}\binom{2(m-i)}{m-i}}{\binom{|Q|}{m}^2} \qquad (6.6)$$

For example, in Figure 6.2, we find the value of $|Q|$ for a given $m$ and $i$. In this case, for $m = 200$ and $i = 10$, we achieve a maximum $p'(i)$ for $|Q| = 3900$.

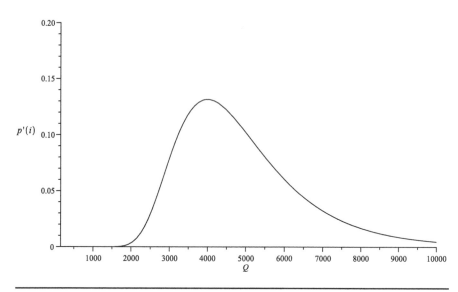

**Figure 6.2.** Key-Pool Set $|Q|$ Selection Based on $p'(i)$ for $m = 200$ and $i = 10$

In general, random key predistribution presents a desirable trade-off between the insecurity of using a single network-wide key and the impractical high memory overhead of using unique pairwise keys. Its main advantage is that it provides much lower memory overhead than the full pairwise keys scheme while being more resilient to node compromise than the single-network-wide key scheme. Furthermore, it is fully distributed and does not require a trusted base station. The main disadvantages to this approach are the probabilistic nature of the scheme, which makes it difficult to provide the guarantee of the initial graph of secure links being connected under nonuniform conditions or sparse deployments. Furthermore, since keys can be shared between a large number of nodes, this class of schemes does not provide very high resilience against node compromise and subsequent exposure of node keys.

### 6.3.6 Random Pairwise Key Scheme

The random pairwise key scheme, proposed by Chan et al. (2003), is a hybrid of the random key-predistribution scheme and the full pairwise key scheme. In the analysis of random key predistribution, it was deduced that as long as any two nodes can form a secure link with at least a probability $p$, the entire network will be connected with secure links with high probability. Based on this observation, Chan et al. note that it is not necessary to perform full pairwise key distribution to achieve a network where any two nodes can find a secure pathway to each other. Instead of preloading $n-1$ unique pairwise keys in each node, the random pairwise key scheme preloads $m \ll n$ unique pairwise keys from each node. The $m$ keys of a key ring are a small, random subset of the $n-1$ possible unique keys that this node could share with the other $n$ nodes in the network. Using the same reasoning as the random key-predistribution scheme, as long as these $m$ keys provide some sufficient probability $p$ of enabling any two neighboring nodes to establish a secure link, the resultant graph of initial secure links will have a high probability of being connected. The remaining links are then established using this initial graph exactly as in the random key-predistribution scheme.

Chan et al. (2003) present a preliminary initial distributed-node-revocation scheme that makes use of the fact that possessing unique pairwise keys allows nodes to perform node-to-node identity authentication.

In their scheme, each of the $m$ nodes that shares a unique pairwise key with the target node (i.e., the node's participants) carries a preloaded vote that it can use to signify a message that the target is compromised. These $m$ votes form a Merkle hash tree with $m$ leaves (Merkle 1980). To vote against the target node, a node performs a network-wide broadcast of its vote (i.e., its leaf in the Merkle hash tree) along with the $\log m$ internal hash values, which will allow the other participants of the target to verify that this leaf value is part of the Merkle hash tree. Once at least $t$ participants of a given target have voted, and the votes have been verified by the other $m$ participants using the Merkle hash tree, all $m$ nodes will erase any pairwise keys shared with the target, thus revoking it from the network.

The random pairwise key scheme inherits both strengths and weaknesses from the full pairwise keys scheme and the random key-distribution scheme. Under the random pairwise keys scheme, nodes captured do not reveal information to the rest of the network, and central revocation can be accomplished by just unicasting to each of the nodes that share keys with the revoked node. It also involves a much lower memory overhead than the full pairwise keys scheme. Unfortunately, like the random key-predistribution schemes, it is probabilistic and cannot be guaranteed to work in nonuniform or sparse deployments.

## 6.3.7 Multispace Key Schemes

This class of schemes is a hybrid between random key predistribution and the $\lambda$-secure $n \times n$ key-establishment schemes. These schemes were first proposed by Du et al. (2003). Recall that in random key predistribution, a key pool is first selected from the universe of possible keys. Each sensor node is then preloaded with a set of keys from the key pool such that any two nodes possess some chosen probability $p$ of sharing enough keys to form a secure link. Multispace key schemes use the same basic notion of random key predistribution but use key spaces, where individual keys are used in random key predistribution. Hence, the key pool is replaced by a pool of key spaces and each node randomly selects a subset of key spaces from the pool of key spaces, such that any two nodes will have some common key space with probability $p$. Each key space represents a unique instance of a different $\lambda$-secure $n \times n$ key-establishment scheme [e.g., Blom's scheme (1984)]. If two nodes possess the same key space, they can then perform the relevant $\lambda$-secure $n \times n$ key-establishment scheme to generate a secure session key.

The main advantage of multispace schemes is that node compromise under these schemes reveals much less information to the adversary than occur with the random key-predistribution schemes. However, they retain the disadvantage of being probabilistic in nature (no guarantee of success in nonuniform or sparse deployments), and, furthermore, they experience the threshold-based sudden security failure mode that is a characteristic of the λ-secure schemes. Other schemes have combined λ-secure schemes with constructions other than random key-space selection. Liu and Ning (2003), in particular, describe a deterministic grid-based construction in which key spaces are used to perform intermediary-based key establishment between nodes.

### 6.3.8 Deterministic Key-Predistribution Schemes

One drawback of the random key-distribution approach is that it does not guarantee success. Lee and Stinson (2005), as well as Camtepe and Yener (2007), propose using combinatorial design techniques to allocate keys to nodes in such a way as to always ensure key sharing between any two nodes. The amount of memory required per node is typically some fractional power of the overall supported network size (e.g., $O(\sqrt{n})$). The main drawback of these schemes is that the same keys are shared between many nodes, leading to weaker resistance to node compromise. Chan et al. (2003) have proposed a deterministic scheme using peer nodes as intermediaries in key establishment with similar memory overheads; compared with the combinatorial design approach, this scheme trades off increased communication cost for greater resistance against node compromise.

### 6.3.9 Public Key Algorithms

Although these algorithms are based on asymmetric key cryptography and are more resource intensive than symmetric key algorithms, they offer better security services, which are much needed and highly advantageous in WSN. As a result, there is motivation to pursue research in developing secure and efficient key-distribution mechanisms suitable in a resource-constraint environment such as WSN. Most of the implementations use RSA or elliptic curve cryptography (ECC).

For example, TinyPK uses the Diffie-Hellman key-exchange technique for key agreement between nodes and is based on the legacy RSA cryptosystem. The main motive of this protocol is to facilitate secure communication between external users and the sensor networks. The external user's identity is established by a certificate authority (CA), where his public key is signed by the CA's private key.

Considering the state of the art in large-number factorization, key-size values are usually set to 1024 bits in RSA as lower values are considerably vulnerable to security attacks. In addition, the public key exponent $e$ is set to 3 and all the resource-intensive operations are carried out on external servers. In this model, resource-abundant devices bear the burden of RSA private key operations, and, hence, the sensor nodes maintain higher energy levels during operations.

## 6.4 ID-Based Key-Distribution Schemes

The concept of identity-based cryptography IBC, first introduced by Shamir in 1984, led to several efficient IBC-based signature schemes (Shamir 1984). However, non-mediated IBC-based encryption (IBE) has proven to be much more challenging, and it is only recently that practical IBE schemes have been found. The first efficient and secure IBE scheme was presented by Boneh and Franklin in 2001, which employs Weil pairing on elliptic curves and is considered more efficient than using regular RSA-based counterparts (Boneh and Franklin 2001). Its security is based on the bilinear Diffie-Hellman problem (BDHP), which is considered secure in the random oracle model (Bellare and Rogaway 1993).

The Boneh-Franklin (BF-IBE) scheme is semantically secure against chosen cipher-text attacks, even when an adversary has the private key of any entities other than the one being attacked. Lynn extended the BF-IBE scheme to provide message authenticity without extra computation cost—i.e., receivers can verify the identity of senders and whether the received messages have already been tampered with, even without resort-ing to digital signatures (Lynn 2002).

Based on the latest advances in IBC and related techniques, in the following sec-tion we explore ID-based key-management schemes to bootstrap secure communi-cations among identifiable peers in wireless ad-hoc networks without a Public Key Infrastructure (PKI) (Section 4.2), CAs, key directories, always-online authorities, or manually arranged pairwise preshared secrets among all involved peers.

*System Setup*

Before an IBC-powered WSN becomes fully functional (i.e., allowing peers to join the system and request keying), an offline Private Key Generator PKG (Section 4.3) first picks a random master key $x \in Z_q$ ($q$ is a prime and $Z$ is an algebraic field), and a bilinear mapping $e : G_1 \times G_1 \rightarrow Z_q$. $e$ is defined on the points of an elliptic curve (as a group $G_1$), and has the following property that for any $P,Q \in G_1$ and for any integer $a$ and $b$,

$$e(aP,bQ) = e(P,bQ)^a = e(aP,Q)^b = e(P,Q)^{ab} \qquad (6.7)$$

The PKG then picks a random generator $P$, and publishes $P$, $xP$, $e$, and four chosen cryptographic hash functions as the public-system parameters. Hash functions are used to hash an arbitrary identity (e.g., any ASCII strings) to a point on the elliptic curve ($H_1$) to achieve security against chosen ciphertext attacks and to encrypt plaintext, respectively. The PKG should keep $x$ secret, and no one else can derive $x$ even when they have both $P$ and $xP$. A lot of offline entities can assume the role of PKG, as long as they can keep the master key secret and extract private keys from the master key for peers joining the system and requesting to be keyed. Once the private key is extracted, a peer has no need to communicate with the PKG (nor to keep the PKG online), unless the peer wants to propose a new identity. Also, the offline PKG can key peers in batches (e.g., only during normal business hours), since peers can receive regular, encrypted information even before they request keying. Compared with an online PKI, the offline PKG has many advantages in wireless ad-hoc networks.

With a PKI, whenever a peer $k$ joins a system, the PKI should verify the binding of the public key of $k$ and its identity and broadcast the authenticated public key to all existing peers, or keep the public key in a central directory for queries from other peers. No matter when another peer $i$ wants to communicate with $k$, $i$ has to obtain both the identity and the public key of $k$, and $i$ should have a way of verifying the public key. Therefore, the complexity in obtaining, verifying, and managing public keys creates considerable overhead in energy-constrained systems that rely on radio technologies to exchange identities, keys, and data.

*Peer Keying*

When a peer $k$ joins an IBC-powered wireless ad-hoc network, $k$ proposes a systemwide unique identity $id_k$ (or the PKG appends a timestamp or sequence number to peer identity). The PKG obtains a corresponding point $Q = H_1(id_k)$ on the elliptic curve by hashing $id_k$ and extracts $k$'s private key $pk_k = xQ$ from the master key $x$. $id_k$ can be the email address of $k$, concatenated with temporal or spatial properties (e.g., a@b.com@ date@site). Identity ownership should be easily verified, e.g., by short-range encounters, when peers are passing by the PKG or by sending a request to confirm email to a@b. com. $pk_k$ is conveyed back to $k$ in a secure, out-of-band side channel (e.g., through the ticketing process at a recreation park); the system parameters are periodically broad-casted by the PKG (e.g., through public announcement). To fight against identity theft or spoofing, the PKG should not extract private keys more than once for the same iden-tity, even when claimed by the same entity; instead, by using timestamp or Universal Unique ID, the entire identity space is always collision-free and forward-secure.

The security of the entire system relies on the master key $x$ kept by the PKG, since the private key of all peers in IBC-based wireless ad-hoc networks can be derived from $x$. To reduce the risk of total exposure, even if the PKG is compromised, and to address the concern of key escrow for peers with a new PKG, $x$ can be distributed in a $t$–of–$n$ manner to a group of $n$ PKGs by applying threshold-cryptography techniques.

With threshold cryptography, $k$ thereby derives $pk_k$ alone by combining $pk_k^t$ obtained from any $t$PKG$_t$. Unless there are more than $t$ unknowingly compromised or bogus PKGs, the secrecy of all peers and their private keys is still preserved.

For simplicity, we focus here on keying with a single PKG; our schemes can be extended for $t$–of–$n$ or hierarchical PKGs, as well.

## 6.4.1 ID-Based Bilinear Key-Predistribution Scheme for Distributed Sensor Networks

### 6.4.1.1 Overview of Matsumoto-Imai's Key-Predistribution Scheme

This model proposes a linear key-predistribution scheme for key sharing among $n$ nodes, and subsequent sections discuss a similar model using symmetric bilinear mapping.

Let us briefly review Matsumoto-Imai's scheme (Matsumoto and Imai 1988). Let $q$ be a prime power and let $m$, $l$ be positive integers. Let $\Psi = GF(q)$ and $\Psi^m = \{x \mid x = [x_1 \, x_2 \, ... x_m], x_i \in \Psi, i = 1, 2, ..., m\}$. Let $\Upsilon$ be a set such that each entity in this set be an identity $d_i$ and $d_i \neq d_j \, \forall i \neq j$. Let a one-way algorithm $\Gamma$ exist that implements an injection from $\Upsilon$ to $\Psi^m$.

*Setup:* The key-setup server selects $l$ $(m, m)$ symmetric matrices $M^\tau s$, where $\tau = 1, ..., l$ over $\Psi$ randomly and independently from other entities. The secret key-sharing function $\Phi_i$ is

$$\Phi_i(\omega) = \phi_i \Gamma(\omega)^T \text{ where } \omega \in \Upsilon \qquad (6.8)$$

for each $d_i \in \Upsilon$. Furthermore, $\Gamma(\omega)^T$ is the transpose of $\Gamma(\omega)$ and $\phi_i$ is an $(l, m)$ matrix defined as

$$\phi_i^T = [M_1 \Gamma(d_i)^T, ..., M_l \Gamma(d_i)^T] \qquad (6.9)$$

Each unit receives its own $\phi_i$ from the server. For example, if nodes A and B want to establish a pairwise cryptographic key between each other, A and B would compute $\Phi_A(y_B)$ and $\Phi_B(y_A)$, respectively, such that there are l-vectors over $\Psi$ that are equal.

Let there be a maximum of $N$ sensor nodes, where each of the $N$ nodes have a unique identity ranging from 1 to $N$. Let each of these identities be represented by an $m$ bits ( $m = \log_2(N)$ ).

*Key predistribution:* The key generator generates $l(m \times m)$ symmetric matrices $M^\tau s$, where $\tau = 1, ..., l$ over the finite field $GF(2)$ and $M^\tau s$ is the master secret. It then computes $\Phi_i$ for each node $S_i$ by computing $\phi_i^\tau = d_i M^\tau$, where $\tau = 1, ..., l$, and generates $\Phi_i$, as shown below:

$$\Phi_i = \begin{bmatrix} \Phi_i^1 \\ \Phi_i^2 \\ ... \\ \Phi_i^l \end{bmatrix} \text{ where } d_i \text{ is the } m\text{-dimensional vector and } i = 1, ..., N \qquad (6.10)$$

*Key establishment:* On deployment of the sensor nodes, let nodes $S_i$ and $S_j$ be two neighboring nodes and be one among the $N$ nodes that would like to establish a pairwise key. These nodes could learn their identities $d_i$ and $d_j$ through a broadcast mechanism. On discovering the identities, the node $S_i$ computes a link key as $K_{ij}^\tau = \Phi_i^\tau d_j^T$, where $\tau = 1, ..., l$ ($K_{ij}^\tau$ represents the $\tau$th bit of the possible pairwise link key between the two communicating nodes). Similarly, $S_j$ constructs the link key $K_{ji}^\tau$. In addition, MAC is used for verifying if the same key was generated at both ends.

## 6.4.2 Key Maintenance

In identity-based schemes, the public key of a peer is exactly its identity or a known transformation of its identity. Hence, a peer can receive regular information encrypted with its identity from other peers even before the peer has obtained its private key from the PKG. This unique feature allows asynchronous communications in wireless ad-hoc networks, where autonomous peers can be in active, idle, or sleep state periodically, without global synchronization to conserve energy. Also, this feature reduces the cost of operating the offline PKG, since peers can request keying in batches only after they are actively and willingly involved in receiving information from other peers and when the PKG goes online according to its own schedule. In contrast, in Symmetric Key Cryptography (SKC) or regular Public Key Cryptography (PKC) systems, peers have to establish pairwise shared-keys or obtain public and private keys prior to any secure communications happening—i.e., keying is always mandatory and proactive for all peers, even if they eventually have no secure communications throughout the validity of their keys in these systems. Once a peer obtains its private key, which is extracted from its identity and the system parameters, the peer can decrypt received information encrypted with its identity, authenticate itself to other peers, and sign outgoing messages. Also, peers can bootstrap shared-keys or derive session keys from their identity-based private keys for symmetric security procedures. Once bootstrapped, symmetric procedures have much less overhead than their asymmetric counterparts. Depending on the definition of peer identity, a peer, as well as the PKG, can determine the lifetime of its private key. For example, a peer can present the same identity (e.g., username) to systems with different parameters (i.e., the peer will have different private keys in different systems); therefore, even if its private key is compromised in one system, the information exposure is confined to that system. A peer can, also, present an ephemeral identity (e.g., user@time). Even if its private key is compromised at a certain time, the peer can request a new private key with a partially updated identity in the time portion, without totally losing its identity or forcedly leaving the system. When necessary, a peer can proactively refresh its identity (e.g., user@date) with the PKG and remain forward-secure even if its current private key is captured and compromised by adversaries. To deal with an unknown PKG, a peer can propose a temporary identity (e.g., user@site) to a newly encountered system, while maintaining credentials with other well-known systems. As we mentioned, a peer can request keying with multiple or hierarchical PKGs to reduce its exposure due to compromised PKGs and to ease its concern about key escrow by untrusted PKGs.

The PKG, on the other hand, can also control the validity of peer identities and extracted private keys. For example, a peer should have a way of proving its identity ownership (e.g., a@b.com) or accept assigned identities (e.g., prepaid personal identification number [PIN]). A peer is uniquely identified by its identity, which can be both time and location invariant within the system. No matter how the peer changes its location and status in the system, it solely relies on its identity to receive information and communicate with other peers. In addition, its identity is related to its reputation (e.g., cooperativeness in relaying) and wealth (e.g., collected credits for its cooperation) in the system. If a peer is found greedy and always fails to relay for other peers, this fact can be taken into account when the peer is in need of relaying by other peers. If a peer is found malicious, either persistently or opportunistically, the peer can be excluded from the system by identity blacklisting or key expiring (e.g., the PKG enforces an identity upgrade and refuses to key compromised peers). The PKG can have differentiated policies, e.g., extracting keys of user@month for well-established or reputable peers (e.g., a monthly pass to a recreation park) and of user@day for new or ill-behaving peers (e.g., a one-time ticket). Certainly, the PKG can enforce a systemwide rekeying after a long time period by updating the master key and the system parameters, and peers will need to contact the PKG again to extract their new private key. The irreplaceable role of peer identity in wireless ad-hoc networks leads to the promotion of identity-based key-management schemes in these systems. These key-management schemes can effectively and efficiently bootstrap security procedures to ensure the confidentiality, integrity, and authenticity of information exchange among peers. We now consider key-management algorithms using the concept of bilinear maps.

A major breakthrough in key agreement using bilinear maps came when Joux (2000) proposed a three-party single-round key agreement using pairing. It led to a new paradigm in key-agreement schemes, and several key-agreement protocols using pairing were proposed. Here, we briefly review Joux's key-agreement algorithm.

Let us consider three parties, $A, B, C$, with secret keys, $a, b, c \in Z_q^*$, respectively. $A$ computes $aP$ and sends it to $B$ and $C$. Similarly, $B$ and $C$ compute $bP$ and $cP$, respectively. $B$ sends $bP$ to $A$ and $C$. $C$ sends $cP$ to $A$ and $B$.

$A$  computes

$$K_A = e(bP, cP)^a \tag{6.11}$$

$B$  computes

$$K_B = e(aP, cP)^b \tag{6.12}$$

$C$  computes

$$K_c = e(aP, bP)^c \tag{6.13}$$

Using the properties of the bilinear map, we conclude that

$$K_A = K_B = K_c = e(P, P)^{abc} \tag{6.14}$$

*Threshold-Key Cryptography*

To avoid a single-point failure, the secret key is distributed among $n$ parties with the condition that it will require the joint effort of more than $t$ players out of $n$ players to construct the secret. The security of the system is not compromised if $t$ parties out of $n$ turn malicious. The distribution of the secret key among $n$ parties is usually done by a trusted dealer or by running an interactive protocol among all parties (Gemmell 1997).

*Pairing-Based Threshold-Key Cryptography*

We choose a $(t-1)$ degree polynomial function $f(x) = s + a_1 x + \ldots + a_{t-1} x^{t-1}$ for random $a_1, \ldots, a_{t-1} \in Z_q^*$. For $i = 1, 2, \ldots, n$ compute $P_{pub}^{(i)} = f(i)P \in G_1$ and $P_{pub} = sP$.

Let $H_1 : \{0,1\}^* \to G_1$ be a map to point hash function and $H_2 : G_2 \to \{0,1\}^l$ be another hash function. Before requesting the private share, each player can check

$$\sum_{i \in S} L_i P_{pub}^{(i)} = P_{pub} \tag{6.15}$$

for any subset $S \subset \{1, \ldots, n\}$ such that $|S| = t$, where $L_i$ denotes the appropriate Lagrange co-efficient explicitly given by the formula

$$L_i = \prod_{j \in S} \frac{-x_j}{x_i - x_j} \tag{6.16}$$

Given a user's identity $ID \in \{0,1\}^*$, the PKG plays the role of a trusted third party. For $i = 1, \ldots, n$ it delivers $d_{ID_i} = f(i)Q_{ID} \in G_1$ to player $i$. After receiving $d_{ID_i}$, player $i$ checks

$$e(P_{pub}^{(i)}, Q_{ID}) = e(P, d_{ID_i}) \tag{6.17}$$

If the verification fails, the end user will notify the PKG, which then issues a new keying material.

*Encryption*: Given message $m \in \{0,1\}^l$ and the identity $ID$, compute $Q_{ID} = H_1(ID)$. Let us choose a random $r \in Z_q^*$ and set the ciphertext to be

$$C = \{rP, m \oplus H_2(e(P_{pub}, Q_{ID})^r)\} = \{U, V\} \tag{6.18}$$

*Decryption*: When receiving $\{U, V\}$, player $i$ computes his decryption share $e(U, d_{ID})$ and gives it to the recombiner who may be a designated player.

*Recombination*: The recombiner selects a set $S \subset \{1, \ldots, n\}$ of $t$ acceptable share $e(U, d_{ID_i})$ and computes

$$g = \prod_{i \in S} e(U, d_{ID_i})^{L_i} \tag{6.19}$$

Once he has $g$, he recovers the plaintext $m = V \oplus H_2(g)$.
Verification is done as shown below

$$g = e(rP, \sum_{i \in S} L_i d_{ID_i}) = e(rP, sQ_{ID}) = e(P_{pub}, Q_{ID})^r \tag{6.20}$$

To check publicly whether the share of a player is acceptable or not, each player chooses a random $R \in G_1$ and computes $w_1 = e(P, R)$, $w_2 = e(U, R)$, and $h = H_1(e(U, d_{ID_i}), e(P_{pub}, Q_{ID}), w_1, w_2)$. Next, player $i$ computes $V = R + hd_{ID_i} \in G_1$ and joins the tuple $(w_1, w_2, h, V)$ to its share. The other players can check that

$$e(P, V) = e(P, R)e(P_{pub}^{(i)}, Q_{ID})^h \tag{6.21}$$

$$e(U, V) = e(U, R)e(U, d_{ID_i})^h \tag{6.22}$$

The security of the scheme is based on the assumption that the bilinear Diffie-Hellman (BDH) problem is hard.

Sakai et al.'s key-sharing algorithm is noninteractive and can be viewed as a type of "dual-identity-based encryption," where the word "dual" indicates that the identities of both the sender and the recipient are required as input into the encryption and decryption algorithm (Sakai et al. 2000).

Let a PKG have a master key $s$, and it issues private keys to users of the form $sP_y$, where $P_y = H_1(ID_y)$ and $ID_y \in \{0,1\}^*$ is the identity of user $y$. Then users $y$ and $z$ have a shared secret that only they can compute, as shown below:

$$e(sP_y, P_z) = e(P_y, sP_z) \tag{6.23}$$

## 6.5 Cross-Layer Approach to Key Management

As sensor networks pose unique challenges, traditional security techniques used in traditional networks cannot be applied directly, for the following reasons. First, sensor networks are limited in their energy, computation, and communication capabilities. Second, unlike traditional networks, sensor nodes are often deployed in accessible areas, presenting the added risk of physical attack. Finally, sensor networks interact closely with their physical environments and with people, posing new security problems. Consequently, existing security mechanisms are inadequate, and new ideas are needed. Owing to resource limitations on computation, storage, and bandwidth, the following aspects should be carefully considered when designing a security scheme:

1. *Power efficiency.* The energy supply is scarce, and, hence, energy consumption is a primary metric to be considered.
2. *Node density and reliability.* WSNs have to scale to much larger numbers (thousands, hundreds of thousands) of entities than current ad-hoc networks, requiring different, more scalable solutions. Sensor nodes are prone to failures. Unfortunately, existing security designs can address only a small, fixed threshold number of compromised nodes; the security protection completely breaks down when the threshold is exceeded.
3. *Adaptive security.* With numerous combinations of sensing, computing, and communication technology, WSNs are conceivable with very different network densities, from very sparse to very dense deployments. They have to interact

with the environment, and the traffic characteristics can be expected to be very different from other, human-driven forms. Therefore, they require different or at least adaptive security protocols.

4. *Self-configurability.* Also similar to ad-hoc networks, WSNs will most likely be required to self-configure into connected networks. However, the difference in factors such as traffic and energy trade-offs may require new solutions. For example, sensor nodes may have to learn about their geographical position.

5. *Simplicity.* Since sensor nodes are small and energy is scarce, the operating and networking software must be kept orders of magnitude simpler as compared to today's desktop computers.

6. *May not have global ID-like IP address.* This is owing to the fact that the global ID will cause a large amount of overhead due to a large number of sensors.

To effectively address the above issues, it may be advantageous to break with the conventional layering rules for networking software (Chapter 8). Due to the limited capacity of sensor node, we should consider a way to save storage space, decrease computational complexity, and reduce communication overhead for key management. An adaptive key-management scheme must be devised to take into account information such as security level, congestion, location, and remaining energy. To this end, one important task is to derive the overall optimization subject to constraints across multiple protocol layers. The key-management scheme, based on such an optimization algorithm, in turn needs to have different components located at multiple layers to work interactively to deliver the overall optimized performance.

## 6.6 Summary

This chapter presents the state of the art in identity-based key-distribution schemes in WSN. Key management is a core mechanism to ensure the security of applications and network services in WSN. It includes two aspects: key distribution and key revocation. Key management has been extensively studied in legacy networks. However, key management has received little attention in resource-constrained networks such as WSN. In this chapter, we summarize the existing key-distribution mechanisms and then examine new ID-based key-distribution schemes in WSN. Unlike most proposed key-distribution schemes, which are based on symmetric key cryptography, we look into new schemes that belong to asymmetric key cryptography and are applicable in a resource-constrained environment. These schemes are more efficient in a distributed peer-to-peer architecture and simplify the process of revocation. We conclude this chapter with a discussion of the cross-layer approach to key distribution in WSN.

## 6.7 References

Aranha, D., R. Dahab, J. López, and L. Oliveira. Efficient implementation of elliptic curve cryptography in wireless sensors. *Advances in Mathematics of Communications* 4, no. 2 (2010): 169–187.

Bellare, M., and P. Rogaway. Random oracle models are practical: A paradigm for designing efficient protocols. *Proceedings of the 1st ACM Conference on Computer and Communications Security (CCS'93)*. ACM, 1993, 62–73.

Blom, R. An optimal class of symmetric key generation systems. *Advances in Cryptology: Proceedings of Eurocrypt '84*, 1984, 335–338.

Blundo, C., A. D. Santis, A. Herzberg, and S. Kutten. Perfectly-secure key distribution for dynamic conferences. *Advances in Cryptology - Crypto '92* (Berlin: Springer-Verlag), 1992, 471–486.

Boneh, D., and M. Franklin. Identity-based encryption from the Weil pairing. *Lecture Notes in Computer Science* (Springer-Verlag) 2139 (2001): 213–229.

Camtepe, S., and B. Yener. Combinatorial design of key distribution mechanisms for wireless sensor networks. *IEEE Transactions on Networking* (IEEE) 15, no. 2 (2007): 346–358.

Chan, H., Perrig, A., and D. Song. Random key pre-distribution schemes for sensor networks. *Proceedings of the 2003 IEEE Symposium on Security and Privacy*. IEEE, 2003, 197–213.

Du, W., J. Deng, Y. Han, and P. Varshney. A pairwise key pre-distribution scheme for wireless sensor networks. *Proceedings of the 10th ACM Conference on Computer and Communications Security (CCS 2003)*. ACM, 2003, 42–51.

Eschenauer, L., and V. D. Gligor. A key management scheme for distributed sensor networks. *Proceedings of the 9th ACM Conference on Computer and Communication Security*. ACM, 2002.

Gemmell, P. An introduction to threshold cryptography. *CryptoBytes, a technical newsletter of RSA Laboratories* 2, no. 7 (1997): 7–11.

Joux A. A one round protocol for tripartite Diffie-Hellman. *Proceedings of Algorithmic Number Theory Symposium – ANTS IV, Lecture Notes in Computer Science* (Springer-Verlag) 1838 (2000): 385–394.

Lee, J., and D. Stinson. Deterministic key predistribution schemes for distributed sensor networks. *Lecture Notes in Computer Science* (Springer-Verlag) 3357 (2005): 294–307.

Liu, D., and P. Ning. Establishing pairwise keys in distributed sensor networks. *Proceedings of the 10th ACM Conference on Computer and Communications Security (CCS 2003)*. ACM, 2003, 52–61.

Lynn, B. Authenticated identity-based encryption. *Cryptology ePrint Archive*-2002/072, 2002

Matsumoto, T., and H. Imai. On the key predistribution system: A practical solution to the key distribution problem. *CRYPTO '87 A Conference on the Theory and Applications of Cryptographic Techniques on Advances in Cryptology*. Springer-Verlag, 1988, 185–193.

Merkle, R. Protocols for public key cryptosystems. *Proceedings of the IEEE Symposium on Research in Security and Privacy*. IEEE, 1980, 122–134.

Perrig, A., R. Szewczyk, J. Tygar, V. Wen, and D. Culler. SPINS: Security protocols for sensor. *Wireless Networks* 8, no. 5 (2002): 521–534.

Sakai R., K. Ohgishi, and M. Kasahara. Cryptosystems based on pairing. *2000 Symposium on Cryptography and Information Security – SCIS'2000*. Okinawa, Japan, January 2000.

Shamir, A. Identity-based cryptosystems and signature schemes. *Advances in Cryptology: Proceedings of CRYPTO 84, Lecture Notes in Computer Science* (Springer-Verlag) 196 (1984): 47–53.

# Chapter 7

# Routing in Wireless Sensor Networks Using Identity-Based Cryptography

## In This Chapter

- Introduction to routing in WSN
- Routing classification
- Selective-forwarding attack
- ID-based routing in WSN
- Cross-layer approach

## 7.1 Introduction

Routing is one of the most fundamental operations in any network that attempts to ensure the delivery of messages from a source to a selected destination. It is a two-step method that involves the process of discovering a suitable route between the concerned source and its destination, and the forwarding of messages using this discovered route. In traditional networks (IP or 3G networks), routing operations are dedicated to special nodes, such as routers. However, WSN consist of resource-constraint devices operating in an ad-hoc decentralized manner that require all the network operations to be done by these ordinary sensor nodes. Some real-time applications (remote-sensing operations) require the routing protocols to facilitate the timely delivery of messages. However, such applications are too resource intensive in WSN and require routing protocols that can balance the energy consumption of the entire network. Furthermore, the number of nodes operating in a WSN scenario is much larger than conventional networks. Consequently, there is a need for the mass production of low-cost nodes. However, with the increase in the number of sensor nodes to meet the current demand for sensor applications, construction of each node to be tamper resistant would be very expensive. As a result, nodes could be susceptible to a node-capture attack. Hence, routing protocols used in traditional networks cannot be applied directly to a resource-constraint environment such as WSN. Therefore, new arrays of routing protocols have been designed for WSN (Al-Karaki and Kamal 2004).

## 7.2 Routing Classifications in WSN

Routing protocols in WSN can be classified by several criteria, such as: data centricity, location information, network layering and in-network processing, path redundancy, a Quality of Service (QoS) requirement, and network heterogeneity.

### 7.2.1 Data-Centric Communication

Conventional networks such as IP networks use a node-centric routing model in which information is exchanged using a unique addressing scheme (IP version 4 or 6). Based on the route the query took to reach the destined node, each source node independently sends data via the shortest path to the concerned sink node. In contrast, a data-centric

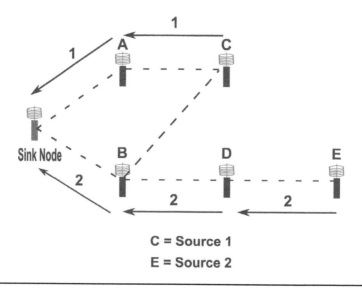

C = Source 1
E = Source 2

**Figure 7.1.** Address-Centric Communication in WSN

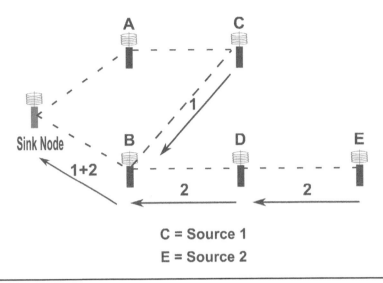

**Figure 7.2.** Data-Centric Communication in WSN

model is more focused on the aggregated data, rather than identifying the exact node's identifiers. Although the request/response scheme is similar in both the models, the sink node or cluster head initiates a request for interested data and the responsible nodes respond with the requested data; they vary in the manner in which the nodes send data back to the sink node or cluster heads. The intermediate routing nodes inspect the data that is being sent to the sink node and perform some form of consolidation operation, such that the sink node receives aggregated data from different sources. Figures 7.1 and 7.2 illustrate the distinction between address-centric and data-centric models in WSN. Figure 7.1 shows the address-centric model in which two sources (nodes C and E) send information to the sink node via the shortest path. Node C sends via node A and node E sends via nodes D and B.

In contrast, Figure 7.2 shows the data-centric model in which data from node C is directly sent to node B for consolidation, and, subsequently, the aggregated data is then sent to the sink node. In cases where node C cannot directly contact node B, an intermediate node closer to node B is responsible for forwarding the data to node B.

On comparing both models, the data-centric model is more energy efficient because only four messages are utilized in sending information from two different sources to the sink node, as opposed to five messages in the address-centric model (Figure 7.1).

## 7.2.2 Location Information

The physical location of a node in a network is an essential metric for designing routing protocols in a WSN. For example, the protocols could be designed for short- or long-range communication, depending on the position of the nodes. In addition, the positions of the nodes influence the design of the forwarding mechanism which, in turn, affects the overall energy consumption of the system. In situations in which routing tables or global knowledge of the network is not required, location-based routing could be useful from a scalability point of view. However, one of the major challenges with such networks is that each node should be aware of its position with reference to the sink node. Additionally, the use of Global Positioning System (GPS)-based chips could weigh heavily on the limited resources available in a sensor-network environment, leading to an increase in the price and energy consumption of the system.

### 7.2.3 Network Layering and In-Network Processing

The architecture of a network could be flat, in the sense that all sensors have the same role. In other words, all sensors forward their sensed data to the sink without necessarily passing through a particular node. A network is said to be *nonlayered* if all sensors form only one group in which the sensors collaborate together to accomplish a common monitoring task. On the other hand, the sensors in a network can be grouped into *clusters*, each of which is managed by a specific sensor called a *cluster head*. These types of networks are considered to be *layered*, and any sensed data should pass through one or more cluster heads before reaching the sink. These cluster heads are supposed to be powerful enough so that they can process the data they receive before sending it to the sink.

All other sensors only need to sense the environment and send their data to the cluster heads for further processing. In some sensing applications, redundancy and correlation exist in the gathered data. Hence, it would be desirable to only transmit more representative data. For example, in monitoring the temperature of a room, the variation in the data within a given region is expected to be small. Thus, the sink is not interested in receiving all the temperature measures, but rather only some of them. This can significantly reduce the communication overhead introduced by data forwarding and improve network performance. In addition, the concept of layering makes a network more scalable and leads to more efficient usage of the energy of sensors, thus extending the network's lifetime.

Extending the network's lifetime is an ultimate goal in the design of a WSN. Given that most of the energy of a sensor is mainly consumed in processing, sensing, and communication, an efficient design approach should take into account these three components of energy consumption. A question that network designers are mostly concerned about is, *How can the lifetime of a network be extended?* To address this problem, several energy-efficient routing and data-dissemination protocols have been proposed that focus on how to forward the data until they reach the sink, regardless of the type of data being transmitted from the source sensors to the sink. Among those protocols, one class does not update the data at the intermediate sensors. That is, each intermediate sensor only acts as a pure data relay without altering any of the data it has received. Another class of protocols introduces the concept of *in-network processing* to handle unnecessary redundancy and correlation contained in the sensed data. In many applications, the data sensed by the sensors has a certain amount of redundancy and correlation. It would be desirable if the sink could only receive relevant data, for faster and better decision making. For this purpose, the sensed data should be processed at intermediate sensors before reaching the sink. The benefit of this in-network processing, such as data fusion, can be seen when vector data rather than scalar data are being transmitted. For example, in an application monitoring the temperature of a room, the sensed data is scalar (i.e., integer or real values). Hence, the cost of data communication is not very high, and the data fusion or aggregation is not as costly. But continuously sending unnecessary and redundant data will consume a huge amount of energy. If a sensing application has to send a large amount of data, for example, images, to the sink for further analysis and processing, it would consume a huge amount of energy. In this case, it would be more beneficial if those images, sensed by different sensors, could be aggregated and only a few of them sent. However, it is also true that processing those images for data fusion requires a considerable amount of energy. Moreover, there will be a delay due to the processing of those images. Therefore, there is a trade-off between data communication and fusion, in these types of information-intensive networks, where the sensed data is not scalar but rather vector.

### 7.2.3.1. Low-Energy Adaptive-Clustering Hierarchy

In terms of the energy efficiency of a WSN, traditional routing protocols based on a flat architecture were not very efficient. Hence, alternative methods were considered in which the clustering approach emerged as a viable option among the existing alternatives. In a cluster-based model, the network is scattered into several clumps or clusters, and each cluster is supervised by a resource-abundant node called a cluster head. The cluster head is in charge of all the necessary coordination between the sensors and acts as a local sink to all the nodes in the cluster. The cluster head could preprocess the aggregated data before forwarding it to the concerned sink node. As a result, a clustering-based approach would be beneficial in a WSN environment since the transmission distance between the ordinary nodes and their respective cluster heads is much smaller, when compared to their respective sink nodes. One such example of a cluster-based protocol that aggregates the sensed data into smaller sizes of data so that only meaningful information is conveyed to all the further forwarding nodes is the Low-Energy Adaptive Clustering Hierarchy (LEACH) (Heinzelman et al. 2002).

LEACH divides a network into several clusters of sensors that are created by using localized coordination and control, so that it not only reduces the amount of data transmitted to the sink but also makes routing and data dissemination more scalable and robust. To conserve the energy levels of ordinary sensor nodes, the selection of cluster heads is not done in a deterministic manner. Instead, it uses a randomized rotational method in which every node in the network has an opportunity to be the cluster head. Consequently, constant depletion of energy levels of individual sensors is significantly reduced. This scheme can be classified into five phases: advertisement, cluster setup, schedule creation, and data transmission.

*Advertisement phase:*

In the beginning, when clusters are being created, the decision about whether a node should be a cluster head is dependent on the percentage of already-existing cluster heads and the number of times an ordinary node has accepted the responsibility of a cluster head. In this model, the assumption is that all nodes in the network begin with the same amount of energy, and that, on taking on the role of a cluster head, almost an equal amount of energy is consumed by each cluster head. Any node $n$ in the cluster would choose a random number between 0 and 1. If this number is less than the threshold $T(n)$, this node would step up to be the cluster head for the current round. Therefore,

$$T(n) = \begin{cases} \dfrac{P}{1 - P\left(r \bmod \dfrac{1}{P}\right)} & \text{if } n \in G \\ 0 & \text{otherwise} \end{cases} \tag{7.1}$$

where $P$ is the desired percentage of cluster heads, $r$ is the current round, and $G$ is the set of nodes that have not been cluster heads in the last $1/P$ rounds. Considering $T(n)$, each node will take on the role of a cluster head at some point within $1/P$ rounds. Let $P$ be the probability that each node will be a cluster head at round 0. Subsequently, these nodes cannot be cluster heads until $1/P$ rounds. Now that we are left with fewer nodes that can take the role of a cluster head, the value of $P$ must be increased. After $(1/P) - 1$ rounds, the threshold values $T(n) = 1$ for any nodes that have not yet assumed the role of a cluster head. In addition, after $1/P$ rounds, all nodes again become eligible to take on the responsibility of a cluster head.

At each round, the selected cluster heads broadcast an advertisement to all the other ordinary nodes in the cluster using the CSMA-MAC protocol and use the same amount of energy for transmission. For this reason, all ordinary nodes should keep their radio in the listening mode to hear the advertisement from the concerned cluster heads. Upon receiving the advertisement message, each ordinary node inspects the received signal strength of the message and finalizes the cluster to which it belongs. Assuming that there is a symmetric propagation channel between the communicating nodes, advertisement messages received with the largest signal strength would indicate that it is closest to that cluster head and would thereby take less transmission energy. In the case of an ordinary node receiving messages with equal signal strength, the node can pick any one of the cluster heads in a randomized manner.

*Cluster set-up phase:*

Upon selecting the cluster heads, each node in the network responds back to its respective cluster head using the CSMA-MAC protocol.

*Schedule-creation phase:*

Upon receiving all the messages from the ordinary nodes, the cluster heads create a TDMA schedule informing each node, by broadcasting the time slots in which they should transmit messages.

*Data-transmission phase:*

Once the ordinary nodes in the cluster have received the TDMA schedule, they are set to begin their usual sensing operation and transmit data at their allocated time slot. To minimize energy dissipation, the radio of each node is turned off until its allocated time. Upon receiving data from all the non–cluster-head nodes, the cluster head can preprocess the data and compress it before sending it to the sink node. In addition, to prevent interference among the nodes, different CDMA codes can be used, and non–cluster-head nodes can decide from the existing list of spreading codes. Subsequently, the cluster heads can inform all the ordinary nodes to transmit using the new spreading code, and they will filter all the messages received using this code.

Despite the fact that LEACH helps the sensors conserve energy in the system, the cluster heads could expend a larger amount of energy when they are positioned farther away from the sink node. In this model, each cluster head sends data directly to the sink, thus presenting one of the main drawbacks with LEACH. A more pragmatic approach is to allow multi-hop data transmission to the sink node through intermediate cluster heads. Consequently, a cluster head does not need to consolidate the processed data from other cluster heads; it only needs to forward the data to the nearest sink node. In addition, this model does not consider the residual energy of the nodes in deciding whether the node should be a cluster head.

### 7.2.3.2 Cluster-Based Energy-Aware Routing

In general, WSN applications can be classified by their functionality (event-based sensing, continuous sensing, and query-based sensing). An event-based sensing application is based on a threshold value. If the sensing parameter value exceeds the threshold, the sensing application in the node is initiated—for example, room temperature measurement. As a result, the energy depletion of the entire sensor network system occurs at a lower rate. For the continuous-sensing application, the nodes are required to continuously sense their surroundings and can be required to transmit data to the sink on a real-time basis (e.g., remote-sensing application). Therefore, such applications demand the continuous sensing of data, which could result in rapid energy depletion of the sensor nodes. In the event of a query-based sensing application, the sink node initiates a query, requesting information from the nodes in a designated region. Upon receiving the query, the sensing application of those designated nodes is activated, and subsequently, an aggregated response is sent to the sink node.

Another approach is a hierarchical, secure, cluster-based, energy-aware routing (SCEAR) scheme (Mallanda et al. 2004). SCEAR proposes an energy-efficient solution by combining a noninteractive identity-based public key cryptography with a symmetric key-based encryption scheme (Maurer and Yacobi 1996; Boneh and Franklin 2001; Khalili et al. 2003). This protocol assumes the following:

- Each entity in the sensor network is static and the nodes are distributed in a random fashion.
- There could be more than one sink node (resource-abundant device), which could act as a gateway to the entire network.

- The deployed sensor nodes are aware of their energy levels. In addition, the nodes also have the capacity to gauge the energy levels of its immediate neighbors.
- A query-based sensing mechanism is used for information retrieval, and the information needed is an attribute of a given geographical area.

SCEAR can be categorized into three phases, namely, initialization, identification, and secure communication.

*Initialization:*

Each node in the network is preloaded with an energy-aware algorithm that helps in self-organizing themselves in each cluster (Chan and Perrig 2004). With the intention of forming a cluster, the elected cluster head broadcasts its identity along with its current status. Depending on the signal strength of the broadcasted message, the nodes in the network decide whether to join the network. The decision to join is dependent on the principle that the greater the strength of the received signal, the lesser is the energy required in sending a message to its cluster head. The interested nodes acknowledge the broadcasted message by sending their identities to their respective cluster heads. For purposes of confidentiality, the node's identity is encrypted with the cluster head's public key (identity of the cluster head). On forming the cluster, each cluster head generates unique symmetric session keys using the RC5 algorithm and sends it to each of its members in the cluster (Rivest 1995). The session key is encrypted using the ordinary node's identity to prevent eavesdropping attacks. Furthermore, the expiration period of the session key is equal to the lifespan of the elected cluster head.

*Identification:*

The moment the cluster heads broadcast their identities (discussed in the *Initialization* phase), the sink nodes closest to the cluster heads also receive this broadcasted message. The sink node responds to this broadcasted message by sending its identity and location information to the nearest cluster head. The identity and location information is always sent through a secure channel by encrypting it using the cluster head's identity. Any pairing- or non–pairing-based identity-based encryption (IBE) algorithm is used for encrypting this information. The neighboring cluster heads then forward this information to their immediate neighbors using the same IBE scheme, and the process continues until all the cluster heads in the network are aware of the sink's identity and location information.

*Secure communication:*

Energy conservation has been the prime requirement of a sensor network, and to adhere to this requirement, SCEAR adopts an event- or query-based sensing functionality. In addition, SCEAR adopts a hierarchical architecture wherein the ordinary sensor nodes encrypt the sensed data with their respective cluster head's public key (identity of the cluster head) before sending the data to the cluster head. Subsequently, the cluster heads aggregate the sensed data using a data-centric routing model and forwards it to the nearest sink node (Section 7.2.1). When a cluster head needs to send information to the nearest sink, the aggregated data is encrypted using the identity (public key) of the sink node, and the header of the packet is encrypted using the public key of the next hop cluster head that is en route to the nearest sink node. The intermediate cluster head decrypts the header, and depending on the destination field in the header, the packet is forwarded to the appropriate neighboring cluster head or the sink node. In case the sink node decides to send a query-based message, it should first locate the identities of the cluster heads belonging to the desired sensing region. To retrieve the identities, the sink sends an ID-request message encrypted with the public key of the immediate next hop neighboring cluster head. The cluster head decrypts the packet and verifies if it is the destination addressed. If not, it is forwarded to the appropriate destination in a way similar to the route taken by the sink that initiated the request. Once the requested packet reaches the destined cluster head, it promptly responds to the query by sending its identity in an encrypted manner.

### 7.2.4  Path Redundancy

The design of WSN should not only consider scalability and energy efficiency but also be robust in nature, which means that a network remains operational despite the occurrence of sensor-node and link failures. The reasons for the failure could be intentional (security attack) or unintentional (defective node or natural calamity). One of the approaches to make the system more robust is to incorporate multipath routing. In short, multipath routing implies the existence of multiple paths (disjoint or partially disjoint) between source node and destination sensors (cluster heads or sink node) (Ganesan et al. 2001).

Although maintaining alternate paths in a routing table introduces some overhead and consumes more energy, multipath routing is an effective technique to improve robustness when link failures occur. Link failures could occur for many different reasons, such as frequent topological changes due to unreliable wireless communication links. Moreover, multipath routing enables recovery from sensor and link failures and provides the necessary resilience to the network at the cost of excessive redundancy.

### 7.2.5  Quality of Service (QoS)

WSN applications have varied requirements that are usually expressed in terms of some metrics, such as delay, fault tolerance, and reliability. For example, real-time applications (video surveillance) are dependent on delay bounds.

Hence, for such applications, the sensed data must reach the sink within a certain time. In addition, a desired property of WSN applications is fault tolerance, which means that a network should remain functional in the event of sensor or link failures. Another desired property is reliability, wherein the aggregated data should be received by the sink as correctly as possible. This would ensure accurate decision making by the sink node. However, metrics such as fault tolerance and reliability necessitate the deployment of additional sensors, yielding additional energy consumption, so that the network can recover swiftly and deliver accurate sensed data to the sink, despite some sensor or link failures. Hence, routing and data-dissemination protocols should consider the trade-offs between fault tolerance, reliability, energy, and delay. Recall that energy is a constraint that should be considered by any routing and data-dissemination protocol to guarantee an efficient usage of the amount of energy available at each sensor.

### 7.2.6 Network Dynamics

Requirements such as limited-energy use (discussed previously) and goals such as mobility have had direct impact on the design decisions of WSN network topology. In theory, a deterministic sensor deployment approach would provide even coverage of the area that has to be sensed. In addition, this approach would require fewer numbers of sensor nodes for accomplishing the required sensing task. However, in a real environment with an uneven terrain, it can be extremely challenging to apply a deterministic sensor deployment strategy. As a result, we are only left with the option of distributing the nodes in a random fashion. Consequently, not all areas of the sensing region are evenly covered by the sensors, leading to a coverage hole. In addition, there is a possibility of not all sensor nodes in the network being connected with each other or even with the sink node. In such situations, mobility plays an important role and becomes the main source of network dynamics that can be used to solve problems. In any sensor network, the aggregated data will be transmitted over some established paths between the source sensors and the cluster heads or sink node. And the establishment of optimal paths depends on whether the sensors are static or mobile. Hence, routing and data-dissemination protocols can be classified based on whether a network is static or dynamic.

In a static network, every node in the network is static—that is, both the sensors and the sink node remain in their fixed positions during their collaborative operation of monitoring a physical environment. Therefore, there is not much overhead required to maintain routes between the sensors and the sink and between the sensors themselves. In particular, the positions of the sensors and the sink can be learned before data exchange, by exchanging some control messages. In certain cases, if the terrain is familiar, node positions can be preconfigured in nodes before deployment. Furthermore, neighbors of a given sensor do not change unless a new sensor has joined the network or an existing sensor has left the network, either by its own will or because of exhaustion of its battery life.

In a mobile network, either the sensors are moving or the sinks or cluster heads are moving. As a result, the routes between the sensors and the sink are changing frequently in such a dynamic environment. Hence, a currently active route could at any time become inactive. This route instability would result in additional overhead and delay in discovering valid routes for data transmission and forwarding. To overcome this drawback, routing algorithms have been proposed in which the ordinary sensors and sinks are designed to be static, whereas certain relay nodes such as cluster heads could be mobile. One such example is the MULES-based architecture (Shah et al. 2003).

In conclusion, the need for mobility in WSN is application dependent. For example, in applications that measure temperature, humidity, sound, or light in an enclosed area, there is no need to have mobile sensors or a mobile sink. However, for monitoring a moving object in a battlefield environment, or in monitoring endangered species, there is a need for mobile sensors in the network, for efficient tracking of the object. In such scenarios, it has been observed that the use of mobile relays helps increase the lifetime of a WSN.

### 7.2.7 Network Heterogeneity

Early research on sensor networks focused on homogenous network architectures (Section 1.3.1). However, recently there has been an increase in the popularity of heterogeneous sensor networks (Section 1.3.2) because of their ability to significantly increase the lifetime and reliability of the system. A usual heterogeneous sensor network consists of a large number of low-cost nodes for the sensing operation and a few resource-abundant nodes that primarily perform data filtering, aggregation, and transport operations. Although heterogeneous networks have gained precedence over homogenous networks, the efficient realization of heterogeneity in a sensor network requires prior systematic planning for placing these heterogeneous resources in a resource-aware manner (Yarvis et al. 2005).

## 7.3 Related Work

The LEACH protocol (discussed in Section 7.2.3.1) has been one of the most promising cluster-based architectures. However, it makes a broad assumption that every ordinary node can easily reach the sink. In addition, ordinary or leaf nodes send the sensed data to their respective cluster heads in order to save energy, which then aggregates the data and sends the aggregation to the sink.

To increase the overall energy efficiency of the sensor network, LEACH randomly rotates cluster heads among all sensor nodes in the network, in a timely fashion. Following the idea of LEACH, a number of protocols have been presented, such as Power-Efficient Gathering in Sensor Information Systems (PEGASIS) (Lindsey and Raghavendra 2002), and Power Efficient and Adaptive Clustering Hierarchy Protocol for wireless sensor networks (PEACH) (Yia et al. 2007), which use a similar concept as LEACH. Because traditional privacy protection cannot solve security vulnerability in WSNs, some security protocols have been proposed for WSNs, e.g., Security Protocols for Sensor Network (SPINS) (Perrig et al. 2002). Adding security to LEACH-like protocols is challenging because its dynamic clustering makes key-distribution solutions for node-to-node trust relationships messy and inadequate. There are some secure routing protocols based on LEACH-like protocols, such as SecLEACH (Oliveira et al. 2007) and grid-based secure LEACH (GS-LEACH) (Banerjee et al. 2007) However, most of them are applying symmetric key-management schemes, which increase the overhead of transmission by establishing pairwise keys. In SecLEACH, for example, nodes that do not share a pairwise key with others in its preloaded key ring cannot participate in any cluster and have to elect themselves as cluster heads. Such a problem is called the orphan node problem, and this problem increases the overhead of transmission. As an evolution of the traditional certificate-based cryptography, ID-based cryptography (IBC) is receiving high regard since its invention (Section 4.3). The main idea is to derive an entity's public key from its identity information (email address, SIP URI, or any biometric information). Recently, the concept of IBC has been increasingly considered in WSN environments (Chapters 5 and 6). In the next section, we look at routing protocols in WSN and the applicability of IBC in WSN routing.

## 7.3.1 Routing Protocols in WSN

Routing in ad-hoc networks has been very challenging due to node mobility. Hence, a routing path established in the beginning between the source and the destination may not exist at a later time interval. Furthermore, in a resource-constraint environment such as WSN, the energy levels of the intermediate nodes must be considered in making routing decisions.

Routing protocols in WSN can be broadly classified into proactive, reactive, hybrid, and location-aware routing protocols (Xiao et al. 2007). In a proactive routing scheme, each node maintains an up-to-date routing table by frequently querying its immediate neighbors for routing information. An example of such a scheme is the Destination Sequenced Distance Vector (DSDV) routing protocol (Perkins and Bhagwat 1994). However, one of the major drawbacks with such schemes is the additional overhead due to frequent routing updates. In contrast, reactive routing involves on-the-fly route establishment and is demand driven. It is based on a request–response model. The initial discovery phase, to find the destined node, could involve flooding, and the response phase establishes the transient active routing path. Examples include Ad-hoc On-Demand Distance Vector (AODV) routing and Dynamic Source Routing (DSR) (Perkins and Royer 1999; Johnson and Maltz 1996).

Various hybrid protocols use the node-discovery method of the proactive routing protocol, along with the on-the-fly routing-path establishment method to produce a hybrid version of the protocol. Zone Routing Protocol (ZRP) is an example of such a hybrid scheme (Haas and Pearlman 2001). In position-aware routing protocols, the nodes select the geographically closest neighboring node when making routing decisions. An example of such a protocol is the Geographic and Energy-Aware Routing (GEAR) protocol (Yu et al. 2001). However, GEAR does not take security into consideration. Most of the security schemes in WSN have focused on symmetric-key cryptography, due to the notion that asymmetric-key cryptography (RSA-based algorithms) was computationally intensive. However, symmetric-key cryptography has major drawbacks with regard to key management, and the security is based upon preshared secret keys. With the successful implementation of pairing-based cryptographic algorithms in WSN, a new platform is provided to implement asymmetric-key cryptographic schemes in WSN (Aranha et al. 2010).

## 7.4  Selective-Forwarding Attack in WSN

Many routing protocols in WSN use a breadth-first spanning-tree algorithm to broadcast routing updates (Newsome et al. 2004; Kuo-Feng et al. 2009). The sink node periodically broadcasts updated routing information to its immediate cluster heads. Then, these cluster heads re-broadcast this information to their immediate neighbors, and the process continues recursively. During this process, each intermediate node makes a note of its parent node, where the parent node is the first node that was able to make contact with its subordinate node and relay the routing information. When all the active nodes are operational, they should send all the sensed data to their parent node. However, this protocol is vulnerable to many attacks.

### 7.4.1  ID-Based Routing in WSN

We now review the ID-based routing method proposed by Lu et al. (2010).

#### 7.4.1.1  Preliminaries

The security of this protocol is based on the Bilinear Diffie-Hellman problem, which, in turn, leads to the fact that the Discrete Logarithm Problem (DLP) for pairings is a *hard* problem.

Let $p$ and $q$ be two large primes and let $E$ be an elliptic curve defined over a finite field $F_p$. Furthermore, let $G_1$ be a $q$-order subgroup of the additive group of points in $E/F_p$, and $G_2$ be a $q$-order subgroup of the multiplicative group in the finite field $F_p$. Let there be a pairing $e: G_1 \times G_1 \rightarrow G_2$, which is a bilinear map satisfying the following properties : bilinear, nondegeneracy, and computability. Examples of such pairings include the Weil and Tate pairing discussed in Chapter 3.

#### 7.4.1.2  Network Architecture and Operation Mechanism

In this model, the network architecture is assumed to be static and homogenous in functionalities and capabilities. The sink is assumed to be reliable and tamper resistant, whereas the sensor nodes are subject to all the attacks mentioned in Chapter 2. In addition, the architecture is hierarchical, wherein the ordinary nodes send data to their respective cluster heads, which, in turn, aggregate and preprocess the data before forwarding it to the sink node. Similar to the LEACH model, this model also assumes a symmetric radio channel between the communicating nodes and has the same energy-dissipation levels. Similar to the LEACH model, the cluster head selection is done in a randomized fashion and is performed in the set-up phase, as shown in Figure 7.3.

In each round, cluster heads are selected in the set-up phase, and in each frame, a sensor node transmits the data to its respective cluster head. In addition, non–cluster

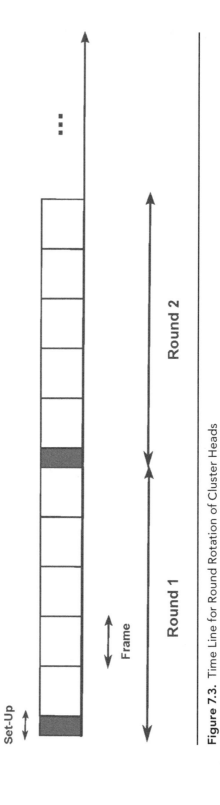

**Figure 7.3.**  Time Line for Round Rotation of Cluster Heads

heads join the cluster using one-hop transmission. As in LEACH, each node determines a random number and compares it with the threshold $T(n)$ in the current round to gauge if it would be elected to be a cluster head.

### 7.4.1.3 Security Vulnerabilities in Cluster-Based Routing Protocols

Similar to any routing protocol in WSN, cluster-based routing protocols (LEACH-like protocols) are susceptible to many security attacks. As these protocols depend on cluster heads, attacks on cluster heads could be more detrimental than on ordinary sensor nodes.

If an attacker is able to compromise and impersonate a cluster head, it can provoke attacks, such as a man-in-the-middle, black hole, or selective-forwarding attack (Figure 7.4), thus compromising the entire network.

For example, a simple impersonation attack leading to a sinkhole attack could totally compromise the entire network in a cluster-based hierarchical architecture (Section 2.2.2 ).

In the traditional Public Key Infrastructure (PKI)-based architecture (Section 4.2), the compromised cluster head in cluster 1 immediately relays information about its revoked key to all of its immediate neighbors. Let us consider the case where a rogue sensor node is one of its immediate neighbors and is a resource-abundant device. This rogue sensor node in cluster 1 will tunnel the revoked key information to its colluder in cluster 2, even before the compromised cluster head in cluster 1 informs its peer in the other cluster. Now, the resource-abundant adversary in cluster 2 can impersonate the cluster head in cluster 1 and generate a new public key as it is aware of the compromised private key. It can reauthenticate to the cluster head in cluster 2 and all the sensor nodes in cluster 2. After proving its authenticity, it can launch a selective-forwarding,

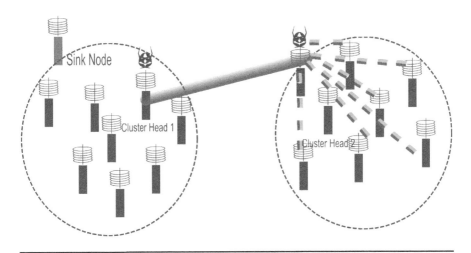

**Figure 7.4.** Selective-Forwarding Attack in WSN

eavesdropping, and black-hole attack. In addition, the attacker can also launch a route-suppression attack by advertising higher energy levels than its compatriots and attract all the data.

Along with active attacks, as discussed above, LEACH-like protocols are susceptible to many passive attacks. Nevertheless, it has been observed that such protocols are more robust against insider attacks than most other types of routing protocols. As cluster heads in LEACH-like protocols are the intermediary nodes to the sink, which change from time to time on a rotational basis, they are perceived to be more robust to node-capture attacks.

### 7.4.1.4 Protocol Initialization

In this ID-based routing scheme, the time taken for messages to travel from cluster heads to sink nodes and vice versa is denoted as $T_i$, and the time for messages to travel from ordinary nodes to their respective cluster heads is $t_i$. The public key is the node's ID concatenated with the $t_i$ (ID + $t_i$). By adopting the ID-based scheme, any sensor node that would want to authenticate itself to another node has to obtain its private key from the concerned Private Key Generator (PKG) (Section 4.3). Because each private key is valid only during the current time interval, sensor nodes have to obtain a denotation of the new time interval to renew the private key at the beginning of a new round. Upon node revocation, the sink needs to broadcast the compromised node IDs to the sensor nodes, and each node stores the revoked IDs within a certain round.

Before deployment, the key-predistribution phase is performed in the set-up phase:

*Set-up phase:*

- Generate public parameters $PP(p, q, E/F_p, G_1, G_2, e)$, as described in Section 7.4.1.1. In addition, let $P \in G_1$.
- Let there be two cryptographic hash functions, such as $H_1$ for the point-mapping hash function, which maps strings to elements in $G_1$, and $H_2$, which maps arbitrary inputs to fixed-length outputs, as shown below:

$$H_1 : \{0,1\}^* \rightarrow G_1^* \tag{7.2}$$

and

$$H_2 : G_2 \rightarrow \{0,1\}^n \text{ for some } n \tag{7.3}$$

- Let $P_{pub} = \tau P$ as a network public key where $\tau \in Z_q^*$ (private key of the PKG)
- Preload each sensor node with the public system parameters $(p, q, E/F_p, G_1, G_2, e, H_1, H_2, P, \tau)$.

### 7.4.1.5 ID-Based Routing Protocol

In this scheme, each round has two phases: a set-up phase and a steady-state phase. Because of the synchronization of time, sensor nodes know when each round starts and ends.

Assume that user $j$ wants to broadcast a message $m$. It first obtains its private key as $d_j = \tau H_1(ID_j \| t_i)$, where $ID_j$ is the ID of node $j$, and $t_i$ is the time-stamp of the time interval in the current round from TDMA (time division multiple access) control. The sensor then picks a random $\alpha \in Z_q^*$ and computes $\theta = e(P,P)^\alpha$. The sensor node further computes

$$c_j = H_2(m \| t_i \| \theta) \tag{7.4}$$

and let

$$\sigma_j = c_j d_j + \alpha P \tag{7.5}$$

where $\langle \sigma_j, c_j \rangle$ is the digital signature on the message $m$. The broadcast message is now concatenated in the form of

$$\langle ID_j, t_i, m, \sigma_j, c_j \rangle \tag{7.6}$$

Upon receiving $\langle ID_j, t_i, m, \sigma_j, c_j \rangle$, each sensor node verifies the authenticity in the following way: It checks the time-stamp of the current time interval $t_i$ and determines whether the received message is fresh. Then, if the time-stamp is correct, the sensor node further computes

$$\theta' = e(\sigma_j, P)e(H_1(ID_j \| t_i), -P_{pub})^{c_j} \tag{7.7}$$

Using the time-stamp of the current time interval $t_i$ and a random $\alpha$ for deriving $\theta'$, the sensor node does the bilinear transformation and mapping from $\theta'$ to $\theta$. We should have the formula below, if the received message is authentic:

$$
\begin{aligned}
\theta' &= e(\sigma_j, P)e(H_1(ID_j \| t_i), -P_{pub})^{c_j}\\
&= e(\sigma_j, P)e(H_1(ID_j \| t_i), -\tau P)^{c_j}\\
&= e(c_j d_j + \alpha P, P)e(H_1(ID_j \| t_i), -\tau P)^{c_j}\\
&= e(c_j d_j + \alpha P, P)e(\tau H_1(ID_j \| t_i), P)^{c_j}\\
&= (e(d_j, P)^{c_j} e(P,P)^\alpha)e(\tau H_1(ID_j \| t_i), P)^{-c_j}\\
&= e(d_j, P)^{c_j} e(P,P)^\alpha e(d_j, P)^{-c_j}\\
&= e(P,P)^\alpha = \theta
\end{aligned}
\tag{7.8}
$$

Then, if $h(m \| t_i \| \theta') = h(m \| t_i \| \theta) = c_j$, which is equal to that in the received message, the sensor node considers the received message authentic and transmits the message to the next hop or user. If the verification above fails, the sensor node considers the message as a bogus or replaced one, even a mistaken one, and ignores it.

The sensor nodes that decide to become cluster heads broadcast an advertisement message (*adv*) along with their digital signature to all sensor nodes in the network.

The nodes that decide to be ordinary nodes pick a cluster head with whom they would want to associate, based on the largest received signal strength of the *adv* messages. Subsequently, the ordinary nodes communicate with the cluster heads by sending a join request (*join*) message. Then, the cluster heads broadcast an allocation message to their cluster members for communication during the steady-state phase, which includes a time slot schedule (*sched*) by TDMA control yet to be concatenated with the digital signature.

During the set-up phase of a new round, the sink node broadcasts its ID and the denotation of the start time of the current round $T_s$, and a nonce $N$ to all sensor nodes (Figure 7.5) Then, the sensor nodes decide whether to become a cluster head for the current round, based on the threshold $T(n)$ [Eq. (7.1)].

Subsequently, the respective cluster heads broadcast their identities, $T_s$, advertisement message (*adv*), signature $\sigma_j$, and $c_j$.

On completion of the set-up phase, the network system goes into the steady-state phase, in which the sensed data is transmitted back from the sensor nodes to its sink (Figure 7.6).

The ordinary sensor node $j$ transmits the data $m$ along with its $ID_j$, $t_i$, $m$, $\sigma_j$, and $c_j$ to its cluster head. Then, each cluster head sensor node collects messages from all of the members in its cluster and aggregates and fuses the data. In the last step, the cluster heads send the aggregated data $F$ along with $T_s$, $\sigma_i$, and $c_i$. The steady-state phase consists of multiple reporting cycles of data transmissions from ordinary nodes to the cluster heads, and is more time consuming.

## 7.4.1.6 Security Analysis

In this model, using an ID-based digital signature provides two advantages: First, it eliminates the use of certificate and auxiliary authentication information. Also, because only the compromised node IDs have to be stored, it requires very little storage space for node revocation, since the length of a user ID is only 2 bytes. However, the disadvantages of this ID-based digital signature are also obvious because of the high-computation cost for bilinear transformation in pairing. Focusing on the resilience to certain attacks in cluster-based WSNs, such as sinkhole attacks, hello flood attacks, and selective-forwarding attacks, the routing protocol works well. All three kinds of attacks are pointed to cluster heads acting as intermediary nodes, because the leaf nodes are not practical targets for attack in a cluster-based structure. Because attackers do not have valid digital signatures to concatenate with broadcast messages for authentication, attackers cannot pretend to be the sink or cluster heads to trigger attacks. Therefore, during both the set-up phase and the steady-state phase, the proposed secure routing is resilient and robust to sinkhole and selective-forwarding attacks. Together with a round rotating mechanism and a digital-signature scheme, the routing protocol is resilient to hello flood attacks to cluster heads because the cluster heads being attacked are capable of ignoring all the communication packets with bogus node IDs or bogus digital signatures.

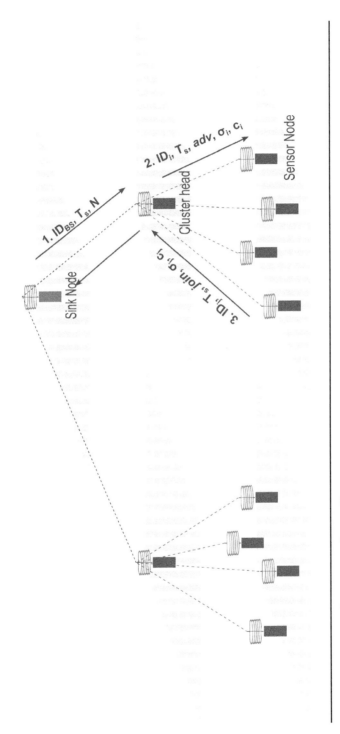

**Figure 7.5.** ID-Based Routing in Setup Phase

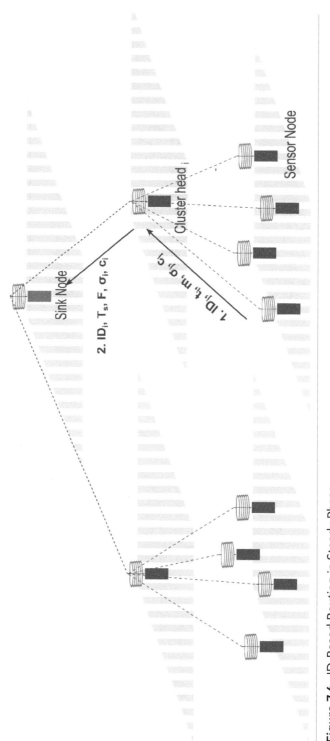

**Figure 7.6.** ID-Based Routing in Steady Phase

*Message size evaluation:*

The total message packet size during the transmissions equals $|ID_j| + |t_i| + |m| + |\sigma_j| + |h(m\|t_i\|\theta)|$, where $|h(M\|t_i\|\theta)|$ is a hash value, which is 20 bytes while SHA-1 is used. $|ID_j| + |t_i|$ are very small, that is, 2 bytes each, and $|m|$ is assumed to be 20 bytes. The total message size of a transmission packet is $44 + |\sigma_j|$ bytes, whereas $|\sigma_j|$ is variable. For example, when using the Tate pairing (Chapter 3), the elliptic curve $E$ is defined over $F_p$, the order $q$ of $G_1$ and $G_2$ is a 160-bit prime. Moreover, $p$ could be a 512-bit prime to achieve a higher level of security than 1024-bit RSA, and $G$ is a $q$-order multiplicative subgroup of the finite field $F_{p^2}$.

## 7.4.2 Cross-Layer Routing

In any communication system, all parties must agree on a set of rules to be used when exchanging messages. For example, rules could include when messages can be sent and the language used for sending these messages. Such an agreement in a communication system is called a *protocol*. Instead of using a single complex protocol that outlines all the details for possible communication, early network designers chose to divide the communication problem into sections and subsections. Furthermore, each section is designed, developed, and tested separately. The main intention of the designers of the layered protocol architecture was to facilitate a functional framework, wherein each layer has a well-defined functional and service definition. Although, this model laid the foundation to support heterogeneous networked systems, it was more applicable to legacy wireless (2G and 3G networks) and IP networks. In addition, the legacy networks focused on QoS and inter- and intracommunication between different networks. On the other hand, WSN focuses on the optimal usage of network resources, thereby increasing the lifespan of the sensor network. It was observed that such a resource-constraint environment would drastically benefit from a cross-layer design in which knowledge shared across different layers of the stack would provide a holistic energy-efficient solution (Shakkottai et al 2003). Akyildiz et al. (2006) proposed one of the early unified cross-layer protocols, which replaced the entire conventional, layered, protocol architecture for WSN. This new protocol efficiently combines the information and the functionalities of the communication layers into a single energy-efficient protocol. The design principle is based on "initiative determination," which embeds intrinsic communication functionalities that are needed for successful communication in WSN. The "initiative determination" concept facilitates the freedom for each node to decide on whether to participate in communication. As a result, the next hop in each communication is not determined in advance. Cross-layer security is discussed in detail in Chapter 8.

## 7.5 Summary

In this chapter, we looked at routing schemes in WSN. We classified routing in WSN into data centricity, location information, network layering and in-network processing, path redundancy, Quality of Service (QoS) requirement, and network heterogeneity. Subsequently, we looked at the applicability of ID-based cryptography in WSN and its efficiency in preventing selective-forwarding attack. In conclusion, we briefly reviewed the concept of cross-layer routing in WSN.

## 7.6 References

Al-Karaki, J. N., and A. E. Kamal. Routing techniques in wireless sensor networks: A survey. *Wireless Communications* (IEEE) 11 (2004): 6–28.

Akyildiz, I. F., M. C. Vuran, and O. B. Akan. A cross-layer protocol for wireless sensor networks. *Proceedings of Information Science and Systems.* 2006.

Aranha, D., R. Dahab, J. López, and L. Oliveira. Efficient implementation of elliptic curve cryptography in wireless sensors. *Advances in Mathematics of Communications* 4, no. 2 (2010): 169–187.

Boneh, D., and M. Franklin. Identity-based encryption from the Weil pairing. *Lecture Notes in Computer Science* (Springer-Verlag) 2139 (2001): 213–229.

Banerjee, P., D. Jacobson, and S. N. Lahiri. Security and performance analysis of a secure clustering protocol for sensor networks. *Network Computing and Application* (IEEE), 2007, 145–152.

Chan, H., and A. Perrig. ACE: An emergent algorithm for highly uniform cluster formation. *Lecture Notes in Computer Science* (Springer) 2920 (2004): 154–171.

Ganesan, D., R. Govindan, S. Shenker, and D. Estrin. Highly-resilient, energy-efficient multi-path routing in wireless sensor networks. *Mobile Computing and Communications Review* (ACM SIGMOBILE) 5, no. 4 (2001): 10–24.

Haas, Z., and M. Pearlman. The performance of query control scheme for the zone routing protocol. *Transactions on Networking* (IEEE), 2001, 427–438.

Heinzelman, W., A. Chandrakasan, and H. Balakrishnan. An application specific protocol architecture for wireless microsensor networks. *Transactions on Wireless Communications* (IEEE) 1, no. 4 (2002): 660–670.

Johnson, D., and D. Maltz. Dynamic source routing. *Mobile Computing* (Kulwer Academic Press), 1996, 153–181.

Khalili, A., J. Katz, and A. Arbaugh. Towards secure key distribution in truly ad-hoc network. *Symposium on Applications and the Internet Workshops.* IEEE, 2003, 342–346.

Kuo-Feng, S., W. Wei-Tong, and C. Wen-Chung. Detecting Sybil attacks in wireless sensor networks using neighboring information. *Computer Networks* (Elsevier) 53, no. 18 (December 2009): 3042–3056.

Lindsey, S., and C. S. Raghavendra. PEGASIS: Power-efficient gathering in sensor information systems. *Proceedings of Aerospace Conference.* IEEE, 2002, 1125–1130.

Lu, H., J. Li, and H. Kameda. A secure routing protocol for cluster-based wireless sensor networks using id-based digital signature. *Global Telecommunications Conference.* IEEE, 2010, 1–5.

Mallanda, C., S. Basavaraju, A. Kulshrestha, R. Kannan, A. Durresi, and S. S. Iyengar. Secure cluster based energy aware routing for wireless sensor networks. *Proceedings of International Conference on Wireless Networks.* CSREA, 2004, 461–466.

Maurer, U. M., and Y. Yacobi. A non-interactive public-key distribution system. *Designs, Codes and Cryptography* (Kluwer) 9, no. 3 (1996): 305–316.

Newsome, J., E. Shi, D. Song, and A. Perrig. The Sybil attack in sensor networks: analysis & defenses. *Third International Symposium on Information Processing in Sensor Networks, IPSN.* IEEE, 2004, 259–268.

Oliveira, L. B., A. Ferreira, M. A. Vilaça, H. C. Wong, M. Bern, R. Dahab, and A. F. F. Loureiro. SecLEACH—On the security of clustered sensor networks. *Signal Processing* (Elsevier) 87, no. 12 (2007): 2882–2895.

Perkins, C., and E. Royer. Ad-hoc on-demand distance vector routing. *Proceedings of the Second IEEE Workshop on Mobile Computing Systems and Applications.* 1999, 90–100.

Perkins, C., and P. Bhagwat. Highly dynamic destination sequenced distance-vector routing for mobile computers. *ACM's Computer Communication Review* (ACM), 1994, 234–244.

Perrig, A., R. Szewczyk, J. Tygar, V. Wen, and D. Culler. SPINS: security protocols for sensor. *Wireless Networks* 8, no. 5 (2002): 521–534.

Rivest, R. L. The RC5 Encryption Algorithm. *Leuven Workshop on Fast Software Encryption.* Springer, 1995, 86–96.

Shah, R. C., S. Roy, S. Jain, and W. Brunette. Data MULEs: modeling a three-tier architecture for sparse sensor networks. *Sensor Network Protocols and Applications.* IEEE, 2003, 30–41.

Shakkottai, S., T. S. Rappaport, and P. C. Karlsson. Cross-layer design for wireless networks. *Communications Magazine* (IEEE) 2003, 74–80.

Xiao, Y., X. Shen, and D. Du. *Wireless network security.* Springer, 2007.

Yarvis, M., N. Kushalnagar, H. Singh, A. Rangarajan, Y. Liu, and S. Singh. Exploiting Heterogeneity in Sensor Networks. *24th Annual Joint Conference of the IEEE Computer and Communications Societies.* IEEE, 2005, 878–890.

Yia, S., J. Heo, Y. Cho, and J. Hong. PEACH: Power-efficient and adaptive clustering hierarchy protocol for wireless sensor networks. *Computer Communications* (Elsevier) 30, no. 14–15 (2007): 2842–2852.

Yu, Y., R. Govindan, and D. Estrin. Geographical and energy aware routing: A recursive data dissemination protocol for wireless sensor networks. Tech Report, UCLA, 2001.

# Chapter 8

# Cross-Layer Security in Wireless Sensor Networks

## In This Chapter

- Introduction to cross-layer design in WSN
- Motivation for cross-layer design
- Cross-layer key distribution

## 8.1 Introduction

Recently, there has been a flurry of cross-layer design schemes being proposed in WSN. As the fusion of secure networking and wireless communication occupies the center stage in sensor networks, the traditional layered protocol architecture upon which most of the networks form their basis is under scrutiny. Although, the layered approach has been repeatedly used in wired networks, it has been argued that the same approach cannot be directly applied in resource-constraint, wireless ad-hoc networks such as WSN. Hence, to combat this approach, security researchers have proposed several cross-layer design schemes in an ad-hoc environment (Shakkottai et al. 2003). Unlike the layering approach, where protocols at each layer are designed independently, cross-layer designs aim at exploiting the dependence between different protocol layers to achieve maximum performance gains. In the current state of the art in the paradigm of cross-layer design schemes in ad-hoc wireless networks, several diverse interpretations exist. One of the main reasons for such varied explanations is that the design effort is largely dominated by researchers who have made independent contributions in designing different layers of the stack. Many of the cross-layer designs depend on other cross-layer designs and hence raise the fundamental question of the coexistence of different cross-layer design proposals. In addition, the question of time synchronization between various cross-layer schemes and the roles each layer of the stack plays is an active area of research. In addition, the wireless medium allows richer modalities of communication than wired networks. For example, nodes can make use of the inherent broadcast nature of the wireless medium and cooperate with each other. Employing modalities such as node cooperation in protocol design also calls for cross-layer design. The goal of designing security solutions with a cross-layer design approach takes us to a new paradigm of security research. The main objective of security solutions in a network is to provide security services such as authentication, integrity, confidentiality, and availability to the users. In wireless ad-hoc networks, due to the unreliable nature of the shared radio medium, attackers can launch varying attacks, ranging from passive reconnaissance attacks to active man-in-the-middle attacks (Chapter 2). Routing in WSN is hop by hop and assumes a trusted, cooperative environment as intermediate nodes act as relays. However, compromised intermediate nodes can launch varying routing attacks, such as blackhole, wormhole, flood rushing, and selective-forwarding attacks (Chapter 2). In this chapter, we review the existing state of the art in the cross-layer design from a security perspective. In addition, as an example, we look at a cross-layer key-distribution mechanism.

## 8.2  Cross-Layer Design in WSN

### 8.2.1  Motivation

As in an Open Systems Interconnect (OSI) model, a layered architecture divides the overall networking task into specific layers, where each layer specifies defined services and direct communication between nonadjacent layers is forbidden. Subsequently, communication between adjacent layers is limited to procedure calls and responses. Such layered architecture facilitates the framework for inter- and intracommunication and sets the ground rules on which protocols can be designed. As a result, a higher layer protocol will only avail the services offered by lower layers and not be concerned about the details involved in implementation of such services. Second, the architecture also implies that the protocol would aim to have standard Application Program Interfaces (API). On the other hand, protocols can be designed by defying the reference layered architecture that could facilitate direct communication between nonadjacent layers. Such violation of layered architecture forms the basis for the cross-layer design architecture. For example, violation of a layered architecture could include creating new interfaces between layers, redefining the layered boundaries, and joint tuning of parameters across different layers. In certain cases, where the violation is too drastic, the original architecture could be replaced completely by the newer cross-layer design. Although, caution should be taken in designing cross-layer design schemes in WSN, there are many incentives to adapting to such schemes that have motivated cross-layer designers to violate the layered architecture. One of the main motivations is the prospect of leveraging on opportunistic communication in wireless links. In the past, protocol designers have faced several challenges in incorporating the legacy layered architecture in the context of wireless communication—for instance, the classic case of misconstruing a Transmission Control Protocol (TCP) packet error on a wireless channel as an indicator of network congestion in the communication channel by the sender of the packet.

TCP is considered to be a reliable, connection-oriented, end-to-end data-transfer protocol. It achieves reliability by error or loss detection and retransmission and is responsible for congestion control in an IP network. In current deployments of IP networks, TCP assumes packet losses and unusual delays in the network as a direct implication of congestion in the network. The receiver on the other end promptly sends cumulative acknowledgments (ACKs) for successfully received segments, which then assist the sender in determining the number of segments that have been received successfully. The sender determines the number of lost packets, either by the arrival of several duplicate cumulative ACKs or by the absence of an ACK for a timeout interval equal to the sum of the smoothed round-trip delay and four times its mean deviation. To reduce congestion in the network, TCP reacts to packet losses, either by retransmitting missing data and simultaneously invoking congestion control by reducing its transmission (congestion) window size or by backing off its repeated retransmissions.. However, when packets are lost for reasons other than congestion, the transmission-rate adaptive behavior of TCP would result in reduced throughput and eventually lead to suboptimal performance of the network. In general, communication over wireless links is often characterized by high bit-error rates due to channel fading, noise, or interference. In such situations, TCP over wireless channels endure considerable throughput degradation and very high interactive delays because the sender misinterprets corruption for congestion. Hence, to alleviate the effects of non–congestion-related losses on TCP over wireless channels, several schemes have been proposed; one of these schemes is Explicit Congestion Notification (ECN) (Floyd 1994).

Whenever the sender generates a TCP packet, it sets the ECN bit to zero. Upon receiving this TCP packet, if the intermediate router detects congestion, it will set the ECN bit to one. Such a TCP packet is said to be marked. This marked packet eventually reaches the destination, and the receiver learns about the new state of the ECN bit in the received TCP packet. Consequently, the receiver informs the sender about the new state of this marked packet to the appropriate sender, which, in turn, adapts its transmission rate depending on the value of the ECN bit. In the current state of the art in TCP deployment, the TCP sender interprets any packet loss to be the end result of congestion. As a result, whenever packet losses occur over a wireless channel, the TCP sender reacts immediately by reducing its packet transmission rate, which results in reduced network throughput.

However, once ECN-enabled TCP is deployed, where the ECN bit can be used to mark packets to indicate congestion, there is a means of differentiating between congestion-related loss and wireless channel–related loss. Thus, the channel need not be smoothed because the ECN mechanism provides a means of explicitly indicating congestion. In Raisinghani and Iyer (2004), it has been analytically shown that in a single-user environment, if packets are marked based solely on congestion information, there is no significant degradation of TCP performance due to the time-varying nature of the wireless channel. A similar idea based on explicit loss notification has been discussed in Zhengrong et al. (n.d.), where simulations indicate improved throughput performance. Such an approach is an example in which a cross-layer view of physical layer information (channel conditions) is used at the network layer to significantly improve network-layer throughput performance. Additionally, the wireless medium offers some new modalities of communication that the layered architectures do not accommodate. For instance, the physical layer can be made capable of receiving multiple packets at the same time. The nodes can also make use of the broadcast nature of the channel and cooperate with one another in involved ways. Making use of such "novel" modes of communication in protocol design also requires violating the layered architectures.

## 8.3 Cross-Layer Design Proposals

In recent times, several cross-layer design schemes have been proposed. Cross-layer feedback optimization could be implemented on the sink node or the sensor nodes. The cross-layer interactions among the layers can be categorized in different ways. For example, Lower to Upper (violation in the flow control from bottom to top), Upper to Lower (violation in the flow control from top to bottom), and Lower and Upper (Figure 8.1). In all these cases, new interfaces will be created between layers. In addition, cross-layer designs can be categorized by the integration of adjacent layers, design coupling without interfaces, and horizontal calibrations (Figures 8.2 and 8.3).

- *Lower to Upper.* The requirement of information from the lower layer to the upper layer at runtime results in the creation of a new interface between these two layers. In this case, the lower layers update necessary information to the appropriate upper layers via the interface. For example, the data-link layer is made aware of the transmit power and the bit error rate information by the

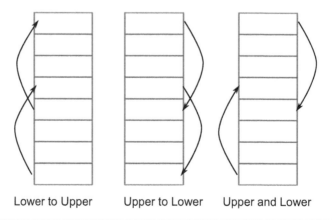

Lower to Upper      Upper to Lower      Upper and Lower

**Figure 8.1.** Cross-Layer Architecture—Vertical Violation

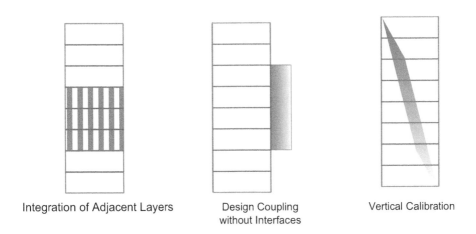

Integration of Adjacent Layers        Design Coupling              Vertical Calibration
                                      without Interfaces

**Figure 8.2.** Cross-Layer Architecture—Integration, Coupling, and Horizontal Calibration

physical layer so that it can adjust its error-correction mechanism. Subsequently, the transport layer can inform the application layer about the TCP packet loss as it would help the upper layer in the stack (application layer) to adjust its transmitting rate. In addition, it should be noted that self-adaptation loops should not be part of a cross-layer design approach, as they do not require new interfaces to be created between the necessary lower and upper layers. For example, in an auto-rate fall-back mechanism for rate selection in a wireless networking environment with multirate physical layers, the Medium Authentication Code (MAC) layer rate selection is dependent on the received acknowledgment that is observable at the MAC layer. Hence, this mechanism would not qualify as a cross-layer design approach as there is no need for creating new interfaces for rate adaption.

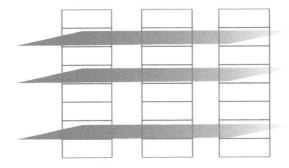

Horizontal Calibration

**Figure 8.3.** Horizontal Calibration

- *Upper to Lower.* The upper layers provide updated information to the necessary lower layers via an interface. For example, if the application layer senses a delay or loss of data, a direct notification to the data-link layer by the application layer would help adapt its error correction mechanism. In addition, delay sensitive packets could be treated with priority. As proposed in Larzon et al. (2002) lower-to-upper information flow is treated as notifications (the lower layer notifies the upper layer about the underlying network condition), whereas the upper-to-lower information flow is treated as hints (upper layers provide hints to the lower layers on the means to process application data).

- *Lower and Upper.* In this case, both the upper and lower layers are at liberty to transmit notifications about their current state and send queries to the other layers. During runtime, layers executing different tasks can collaborate with each other on an iterative loop basis, resulting in a back and forth communication between them. For example, a back and forth information flow between layers is seen in a proposal to solve the multiple access problem for contention-based wireless ad-hoc networks using joint scheduling and suggesting a distributed power-control algorithm for such networks (ElBatt and Ephremides 2004). In addition, direct communication between layers at runtime could indicate the advantage of making the variables at each layer visible to the other layers of the stack. However, one of the disadvantages of this approach would be in managing the shared memory spaces between the layers when variables and internal states are shared between different layers.

- *Integration of Adjacent Layers.* The formation of a super-layer by combining two or more adjacent layers would result in a new cross-layer design scheme. The resulting layer would simply provide the union of the services that were provided by the individual layers. For example, a collaborative design of the data-link and physical layer would suffice to produce a super-layer. From a network security perspective, a super-layer that combines network and data-link layer would help prevent advanced Address Resolution Protocol (ARP) poisoning attacks.

- *Design Coupling without Interfaces.* Coupling two or more layers during the design phase would avoid creating extra interfaces at runtime that could result in a new cross-layer design approach. However, in deployed networks, one of the architectural challenges would be to integrate the coupled layer with already-existing fixed layers.

- Vertical and Horizontal Calibration Across Layers:

  *Vertical calibration.* Vertical calibration refers to the efficient utilization of parameters across different layers of the vertical stack. The parameters set at the application level could dictate terms to the lower layers and vice versa. For example, the transport protocol (TCP or UDP) chosen at the transport layer would assert reliable or unreliable communication and would directly affect the layers below it. Consequently, the joint adjustment at different layers of the vertical stack would result in a more holistic performance of the system than the adjustment of individual parameters.

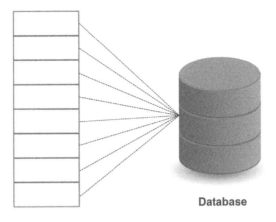

**Figure 8.4.** Shared Database Approach Across Cross Layers

*Horizontal calibration.* Horizontal calibration could be very useful in a resource-constraint environment such WSN. In this case, not only individual parameters pertaining to that layer are taken into consideration, but parameters pertaining to other compatriot layers are also considered. For example, while routing packets, if the network level state of intermediate nodes is taken into consideration, it would be easy to detect nonactive nodes and could subsequently produce an energy-efficient routing protocol. However, challenges do exist in case the participating nodes do not adhere to the same cross-layer approach as the initiating node.

- *Shared Database Across Layers* (Figure 8.4). The use of a common database which emulates a new layer that provides services such as storage/retrieval of information to all the layers is particularly well suited to vertical calibrations across layers. An optimization program can interface with the different layers at once through the shared database. Similarly, new interfaces between the layers can also be realized through the shared database. However, the main challenge to this approach is with the design of the interactions between the different layers and the shared database.

- *New Abstractions.* In this case, cross-layer designs are presented in a completely new way, and each design follows its own set of rules or protocols. For example, using heaps instead of stacks could facilitate rich interactions between protocols and offer greater facility in design and during the run-time environment. On the contrary, such protocols may not interact with the traditional protocols and hence are restricted to closed, proprietary networks.

## 8.4 Cross-Layer Key Distribution

In Section 6.5, we briefly introduced cross-layer key management in WSN. In this section, we look at a cross-layer approach to key establishment that could result in reduced computational cost. We review a cross-layer key establishment protocol using a cluster protocol proposed for WSN (Saglam and Dalkılıç 2009a). This clustering protocol is reviewed in the next section.

### 8.4.1 Clustering Protocol

The clustering protocol proposed in Saglam and Dalkılıç (2009a) is a low-node density (around 7), self-organizing hop-by-hop protocol based on a spanning-tree protocol growing in parallel with an iterative heuristic. The protocol is based on local decisions, and cluster formation is dependent on border members selecting new initiators for the organization of new clusters. During the configuration phase, every node in the network runs the same protocol state machine. There are five states for a sensor node that also define the node types, namely, initiator, node, gateway candidate, floating, and gateway. At the beginning, all the nodes are in floating state and waiting to be configured. A simplified state machine of this clustering protocol is provided in Figure 8.5.

In the floating state, a node may receive a configuration message from one of its neighbors; it then takes corresponding action. When the purpose of this message is to start a new cluster, the state of the receiving node is then changed to initiator, and this initiator shares the maximum available budget (e.g., in terms of energy expenditure) with its floating neighbors. If the intent of the configuration message is for inner cluster extension, then the state is changed to node and it shares the received budget with its floating neighbors. If the available budget with the inner cluster extension message is 1, then the unit changes its state to gateway candidate and starts the gateway decision algorithm defined in Saglam and Dalkılıç (2009a). If this unit finds itself as the best candidate among the other gateway candidates then it changes its state to gateway and tries to select one of its floating neighbors as the initiator of the next cluster to be extended. Otherwise, the state is changed to node and this unit stops its configuration process. This iteration continues until there is no possible way for this spanning tree to extend.

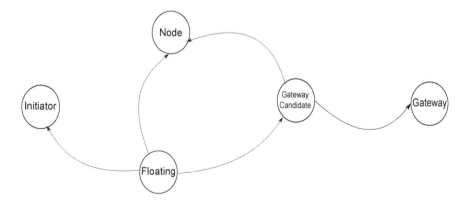

**Figure 8.5.** State Machine for Clustering

## 8.5 Cross-Layer Key Establishment

In this section, we provide the implementation details of the key-establishment protocols embedded within the clustering protocol Saglam and Dalkılıç (2009b). Here, the
physical-node density of the network has been selected as 8, which is the minimum
node density requirement of the clustering protocol to achieve a network connectivity
performance higher than 99%. We first detail the implementation of the basic scheme,
and then provide the ECDH-based key establishment.

### 8.5.1 Basic Scheme

The cross-layer implementation of the basic scheme includes a clustering protocol that
would require additional communication messages in the neighbor-discovery phase. In
the clustering protocol discussed in Section 8.4.1, as soon as a configuration message is
received by a floating sensor node, the neighbor-discovery operation is initiated. During
this phase, states of sensor nodes in the neighborhood are registered and the budgets for
cluster extension are distributed to the floating neighbors. Predistributed key rings are
checked during this phase, and the neighbors sharing common key(s) in their key rings
are used to establish a secure connection (Sections 6.3.4 and 6.3.5).

Key pool size $Q$ and key ring size $m$ of the basic scheme define the probability $p$ of
establishing a secure link between two sensor nodes. Hence, the number of neighbors to
be securely connected is determined by these parameters. However, it should be noted
that changing these parameters affects the cost of storage and transmission of messages.

Transactions for basic scheme implementation in the neighbor-discovery phase of
the clustering protocol are depicted in Figure 8.5.

*Phase I.* In this scheme, any configured sensor node (Alice) starts neighbor discovery
by broadcasting a list of key indexes, $k_A$, included in its key ring (Figure 8.6—*Phase I*).

*Phase II.* This polling message is received by its neighbors $n_i$ ($i = 1, 2, \ldots, d$, where $d$
is the expected node density) and compare $k_A$ with their key rings $k_i$. If there is at least
one key index match, then the corresponding neighbors reply to Alice with the matching index number $idx_i$ (*Phase II*). Hence, the expected number of neighbors finding a
common key after the shared key discovery is $pd$. As a result, there exist $(1-p)d$ neighbors
that cannot find a shared key with $k_A$. These neighbors who do not share a key reply
with a list of their key indexes $k_i$ ($i = 1, 2, \ldots, (1-p)d$), which are subsequently used in
the path key discovery phase.

*Phase III.* In this phase, Alice is aware of the nodes that share a key, and depending on
the probability $p$, if there are neighbors that cannot establish a link key, Alice initiates
the path key discovery process via its $pd$ secure neighbors. Alice broadcasts the list of key
indexes $k_i$ ($i = 1, 2, \ldots, (1-p)d$) received in the shared key-discovery phase together with
the corresponding node identities $id_j$.

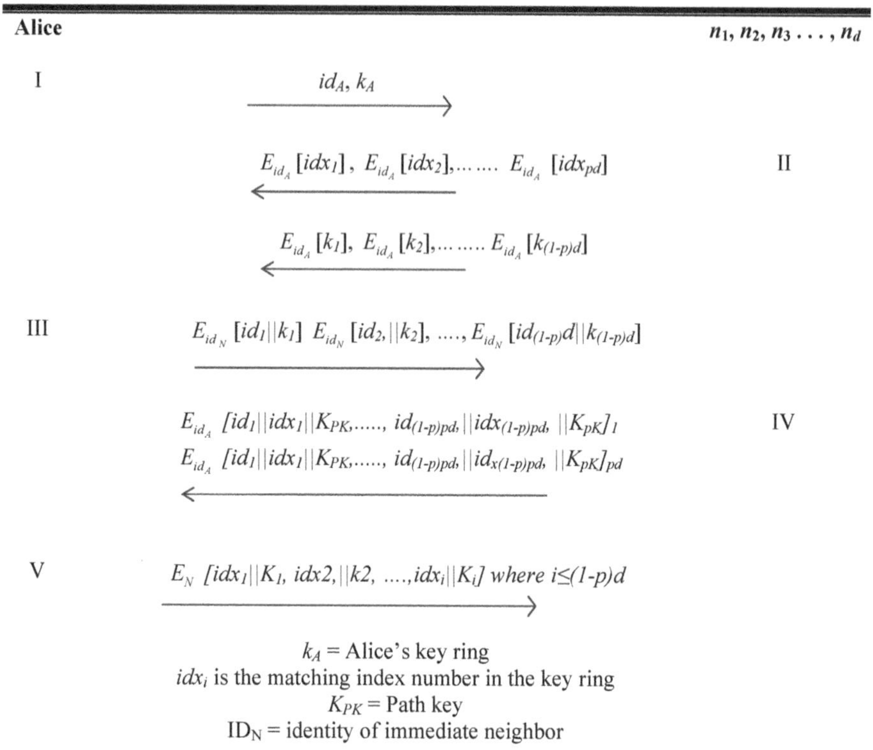

| Alice | | $n_1, n_2, n_3 \ldots, n_d$ |
|---|---|---|
| I | $id_A, k_A$ <br> $\longrightarrow$ | |
| | $E_{id_A}[idx_1], E_{id_A}[idx_2], \ldots\ldots E_{id_A}[idx_{pd}]$ <br> $\longleftarrow$ | II |
| | $E_{id_A}[k_1], E_{id_A}[k_2], \ldots\ldots E_{id_A}[k_{(1-p)d}]$ <br> $\longleftarrow$ | |
| III | $E_{id_N}[id_1\|\|k_1]\ E_{id_N}[id_2,\|\|k_2], \ldots., E_{id_N}[id_{(1-p)d}\|\|k_{(1-p)d}]$ <br> $\longrightarrow$ | |
| | $E_{id_A}[id_1\|\|idx_1\|\|K_{PK},\ldots., id_{(1-p)pd}\|\|idx_{(1-p)pd},\|\|K_{pK}]_1$ <br> $E_{id_A}[id_1\|\|idx_1\|\|K_{PK},\ldots., id_{(1-p)pd}\|\|idx_{x(1-p)pd},\|\|K_{pK}]_{pd}$ <br> $\longleftarrow$ | IV |
| V | $E_N[idx_1\|\|K_1, idx2,\|\|k2, \ldots.,idx_i\|\|K_i]$ where $i{\leq}(1-p)d$ <br> $\longrightarrow$ | |

$k_A$ = Alice's key ring
$idx_i$ is the matching index number in the key ring
$K_{PK}$ = Path key
$ID_N$ = identity of immediate neighbor

**Figure 8.6.** Message Exchange for the Basic Scheme

*Phase IV.* In *Phase IV*, each of the *pd* secure neighbors compare their key indexes with the list provided by Alice to find a common path key. Each discovered path key ($K_{PK}$) is sent to Alice by encrypting it with the link key. As a result, there could be multiple path keys $K_{PK}$ for the distant neighbor ni.

*Phase V.* In *Phase V*, upon receiving the path key, Alice selects only one of them ($K_i$), and the path key is sent to the corresponding neighbor with its key index number $idx_i$, where $i \leq (1-p)d$. After this phase, if there are neighbors that still cannot establish a path key then Alice starts a new path key discovery iteration starting from *Phase III*. New iteration also includes the last added neighbors in the previous path key discovery as the shared key members. This loop terminates when there is no new neighbor added after the last path key discovery process.

In this scheme, the links to be secured are defined by the clustering protocol discussed in Section 8.4.1. Securing all the physical neighbors for each sensor node would give the highest network configuration performance. However, the communication and computation costs are higher in this case. Therefore, only the required neighbors that are in a floating state can be selected for establishing secure links while extending the clusters. This reduction can improve the communication costs and hence the computation costs because of the reduced MAC computation. In this case, three different methods for the selection of links to be secured have been proposed—proactive, reactive, and straight. With the proactive selection, attempts are made to secure the links with all the physical neighbors in the shared key and path key discovery phases. With the reactive selection, attempts are made to secure only the links with floating neighbors in shared key discovery. If all of them cannot be secured, then the path key discovery process is applied for the remaining floating neighbors, just one time. Straight selection only considers the neighbors having shared keys discovered in the shared key discovery phase, and path key discovery does not run in this mode.

### 8.5.2 Pairwise Key Setup with ECDH

The cross-layer implementation of ECDH key exchange with the clustering protocol does not require additional messaging for the clustering protocol. The implementation is simple since for a pair of sensor nodes the transaction only requires the exchange of public keys $Pu$ (identity in case of ID-based cryptography) to calculate the shared key for that link. $A$ first broadcasts the polling message for neighbor discovery, which also includes its public key $Pu_A$. Then all the receiving neighbors $n_i$ $(i = 1, 2, \ldots, d)$ include their public keys $Pu_i$ in their reply messages. The bulk of the cost of this operation is the point multiplication of the neighbor public key with the private key to calculate the link key for each neighbor sensor node.

We assume that the public keys are calculated offline and loaded to sensor nodes before deployment. This operation saves the one-point multiplication operation in each sensor node needed for calculating the public key from the private key. Because the links to be secured are controlled by the clustering protocol, there may be no need for $A$ to establish a secure link with all of its neighbors for acceptable network connectivity performance. Here, we can apply reactive and proactive link selection methods for ECDH, as well. For the proactive selection, attempt is made to secure all the possible links among the neighbors, which will provide the same connectivity performance of the clustering protocol without security implementation. With the reactive selection, attempt is only made to secure the links with floating neighbors. Because the link key probability is 1 for the ECDH, if the network density is selected as low as possible then the computational cost could be reduced accordingly.

## 8.6 Open Challenges

In recent times, we have seen the emergence of cross-layer proposals in WSN. However, open challenges exist in the smooth functioning of a few cross-layer design schemes. Since cross-layer schemes are designed by individuals with specialized skills, the coexistence of different cross-layer proposals with one another in a real-time environment is considered to be an open challenge. In addition, care should be taken that a given cross-layer design scheme won't stifle innovation in the future. In the following sections, we discuss some of the challenges in detail.

### 8.6.1 Cross-Layer Couplings

Although many cross-layer proposals exist that are disparate in nature, identifying proposals that are pertinent to a particular task has been a challenge. For example, the design of a cross-layer scheme between a network and the MAC layers of a WSN protocol stack should involve a cost-benefit analysis of different cross-layer design proposals with reference to complexity versus performance. In addition, adequate security should be considered while selecting a cross-layer proposal. Furthermore, design proposals should incorporate explicit notifications by new interfaces to the transport layer so that they improve end-to-end performance and make use of channel knowledge at the MAC layer that would allow opportunistic usage of the channel and improve performance. In addition, energy-, delay-, and security-related issues need to be handled across the layers in a holistic manner. This would require comparative, quantitative study of the different cross-layer design proposals and is currently an open challenge for the cross-layer community.

### 8.6.2 Coexistence of Cross-Layer Design Proposals

One of the prime challenges with cross-layer security is with the coexistence of different cross-layer security schemes. For example, the MAC layer in the WSN stack responds to variations in the channel by adjusting the data rate. Is it then possible to adjust the frame length at the link layer to provide further help? How will an overriding control from the transport layer, attempting to control the link-layer parameters, interact with these adaptation loops?

The question of the coexistence of cross-layer design ideas is pertinent when it comes to determining whether some cross-layer design proposals can stifle further innovation. Let us say that the physical and link layers are optimized for a certain performance metric in a cross-layer design scheme. If this scheme is deployed first, can other schemes that also rely on some (other) cross-layer couplings, or those that assume no coupling between the link and physical layers, be deployed, as well, at a later time? Apart from presenting new cross-layer design proposals, designers need to start establishing which other cross-layer design interactions may or may not be employed together with their proposal.

### 8.6.3 Particular Cross-Layer Design

The network conditions in a wireless network are usually time-varying. In such a situation, one of the stated motivations behind cross-layer design is to achieve the network equivalent of impedance matching. The idea is to make the protocol stack responsive to variations in the underlying network conditions so that an optimal operating point is always maintained. The pursuit of achieving such optimal operation presents two complementary challenges. First, designers need to establish the network conditions under which the proposed cross-layer designs will result in performance improvements. Therefore, research should be conducted to illustrate how a cross-layer design involving an iterative optimization of throughput and power leads to a loss in performance under a certain uncontrolled network condition. Second, efficient mechanisms to make a timely and accurate assessment of the state of the network needs to be built into the stack, and the corresponding overheads must be taken into account.

### 8.6.4 Standardization of Interfaces

The one thing layering achieved was to present standardized boundaries and interfaces between modules of the system—the protocol layers. Now that the layered architecture is being violated in different ways, finding the new reference architecture becomes a challenge. As a result, the question to ask is, What should the boundaries be between modules? In addition, Should we stick to traditional layer boundaries and determine the new interfaces from there, or should we look at completely new boundaries, or a combination thereof? What should the interfaces between modules look like? Addressing this challenge requires greater synergy between the performance viewpoint and implementation concerns than is seen in the literature today. In general, organization of the modules (layers or otherwise) and the interfaces between them determines how efficiently information can be shared between them and what kinds of overheads and delays occur? This, in turn, determines how effective cross-layer design proposals, which rely on sharing dynamic information between the modules, can be. Hence, proposers of cross-layer design relying on back-and-forth information flow between layers or dynamic vertical calibrations need to consider the impact of delays in the retrieval/updating of information on protocol performance. They also need to quantify the overheads associated with their cross-layer design proposals.

### 8.6.5 Physical Layer in Cross-Layer Proposals

In wired networks, the role of the physical layer has been rather small: sending and receiving packets as initiated by the higher layers. Advances in signal processing at the physical layer can allow it to play a bigger role in wireless networks. Hence, the question of how much of a role the physical layer should play. This is relevant to the cross-layer design effort because first, layered architectures like the OSI reference model do not allow much of a role for the physical layer besides providing a bit pipe, and second, enhancements in the physical layer will have to be balanced by corresponding changes to the higher layers. Hence, figuring out the role to be played by the physical layer is an important question. Cross-layer designs relying on advanced signal processing at the physical layer can be an interesting research framework for the future.

### 8.6.6 Ideal Model for Secure Communication

The last open challenge relates to the communication model assumed. Wired networks, by their very nature, are essentially a collection of well-defined point-to-point communication links. The same cannot be said about wireless networks because the wireless medium is inherently broadcast, and there is no clear-cut concept of a communication link in wireless networks. This gives rise to a fundamental question of whether it still makes sense to "create" links in a wireless network. Some recent work has made use of the inherent broadcast nature of the wireless medium to come up with innovative communication schemes for wireless networks. For example, exploiting the broadcast nature of the wireless medium can allow transmission schemes that rely on cooperation between the communication nodes.

These new modalities that the wireless medium offers cannot be accommodated in layered architectures and, hence, inevitably require some degree of architecture violation—in other words, cross-layer design. They represent new opportunities for the cross-layer design effort. Hence, we believe that while proposing cross-layer design proposals to address the issues raised by wireless links, the designers should also keep an eye out for new opportunities created by wireless communication networks.

## 8.7 Summary

In this chapter we reviewed the cross-layer design approach in WSN, as well as different interpretations of the cross-layer design schemes and their representative cross-layer design proposals. In addition, we looked at initial ideas on how cross-layer interactions may be implemented and, as an example, looked at a cross-layer key-distribution scheme in WSN.

## 8.8 References

ElBatt, T., and A. Ephremides. Joint scheduling and power control for wireless ad hoc networks. *IEEE Transactions on Wireless Communications* 3, no. 1 (2004): 74–85.

Floyd, S. TCP and explicit congestion notification. *SIGCOMM Computer Communication Review* 24, no. 5 (1994): 8–23.

Larzon, L.-A., U. Bodin, and O. Schelen. Hints and notifications. *Wireless Communications and Networking Conference.* IEEE, 2002, 635–641.

Raisinghani, V. T., and S. Iyer. Cross-layer design optimizations in wireless protocol stacks. *Computer Communication* 27 (2004 ): 720–724.

Saglam, O., and M. E. Dalkılıç. A self organizing multihop clustering protocol for wireless sensor networks. *Fifth International Conference on Mobile Ad-hoc and Sensor Networks.* IEEE, 2009a, 33–40.

Saglam, O., and M. E. Dalkılıç. Cross layer implementation of key establishment protocols in wireless sensor networks. *International Symposium on Computer and Information Sciences.* IEEE, 2009b, 346–351.

Shakkottai, S., T. S. Rappaport, and P. C. Karlsson. Cross-layer design for wireless networks. *Communications Magazine* . IEEE, 41, no. 10 (2003): 74–80.

Zhengrong, Ji, Y. Yang, Z. Junlan, T. Mineo, and R. Bagrodia. Exploiting medium access diversity in rate adaptive wireless LANs. *MobiCom '04.* ACM, n.d., 345–359.

# Chapter 9

# Implementation of Elliptic Curve Cryptography in Wireless Sensor Networks

### In This Chapter

- Design principles for the implementation of ECC on WSN
- Selection of ECC security parameters
- Various sensor platforms
- Optimized algorithms for finite field arithmetic for WSN

## 9.1 Introduction

The invention of Public Key Cryptography (PKC) has revolutionized the construction of cryptographic systems and provided a secure foundation for the implementation of cryptographic theory in real-world applications. Efficient computation of PKC in sensor nodes has been extensively investigated by researchers. However, implementing cryptographic operations in devices with power, space, and time constraints limits the sensor's ability to handle the additional computations required by cryptographic operations. PKC is indeed shown to be feasible in WSNs by using Elliptic Curve Cryptography (ECC). However, the resource-constraint nature of WSN has imposed serious challenges to the efficient computation of cryptographic schemes on such networks. In this chapter, we review optimized techniques in prime and binary fields for different arithmetic operations on different sensor platforms.

Until a decade ago, due to the resource-constraint nature of WSN, security goals such as authentication, confidentiality, and integrity protection were accomplished using symmetric techniques. In recent times, however, ECC has emerged as a potential alternative to conventional public key cryptographic techniques such as RSA, Digital Signature Algorithm (DSA), and Diffie-Hellman (DH) (Section 3.6). More specifically, ECC provides equivalent security with a smaller key size, which results in reduced computational time and lower memory requirements. The primary operation underlying RSA is modular exponentiation in integer rings and its security is based on the difficulty in factoring larger integers. Conversely, ECC operates on groups of points over elliptic curves, and its security is based on the hardness of the Elliptic Curve Discrete Logarithm Problem (ECDLP) (Section 3.7). Whereas subexponential algorithms solve the integer factorization problem, only exponential algorithms are known to solve the ECDLP. Hence, ECC achieves higher computational efficiency for the same level of security as RSA with a smaller key size. Consequently, the motivation for increasingly efficient ECC-based algorithms and its implementation has increased in a resource-constraint environment such as WSN.

## 9.2 Point Coordinate System

Many different coordinate systems can represent points on elliptic curves. However, the selection of the point coordinate system has a significant influence on the performance of the elliptic curve arithmetic operations. For instance, point multiplication operations can be expedited by using efficient representation of elliptic curve points. For instance, in ECC, many coordinate systems such as Affine, Projective, Jacobian, Lopez-Dahab, and mixed coordinates can be used to represent points.

*Affine coordinates*: These are standard $(x, y)$ coordinates commonly used in mathematics.

*Projective coordinates*: These may be used in some cases to accelerate certain arithmetic operations. The point $(X, Y, Z)$ on $E$ corresponds to the affine point $(X/Z, Y/Z)$ when $Z \neq 0$ and to the point at infinity $\mathcal{O} = (0, 1, 0)$. It has been shown that projective coordinate systems are more efficient than affine systems when calculating point multiplication.

*Jacobian coordinates*: This system improves the timings for basic arithmetic operations on a curve. The point $(X, Y, Z)$ on $E$ in Jacobian coordinates corresponds to the affine point $(X/Z^2, Y/Z^3)$ when $Z \neq 0$ and to the point at infinity $\mathcal{O} = (1, 1, 0)$. In general for Jacobian coordinates, doublings are faster and additions slower than for the projective coordinates.

*Lopez-Dahab coordinates*: These are designed for curves defined over fields of characteristic two. A point $(X, Y, Z)$ on an elliptic curve in Lopez-Dahab coordinates is equal to the point ($X/Z$, $Y/Z2$) in affine coordinates where $Z \neq 0$. The point at infinity corresponds to $\mathcal{O} = (1, 0, 0)$. These coordinates are the optimal choice when calculating point multiplication on binary curves.

Several software implementations of ECC in WSN have been reported (Gura et al. 2004; Malan et al. 2004; Eberle et al. 2005; Yan and Shi 2006; Szczechowiak et al. 2008). The advantages of software implementations include ease of use, ease of upgrade, portability, low development cost, and flexibility. However, in comparison to hardware implementations, software implementations have some disadvantages, such as a limited ability to protect private keys from disclosure. In the earlier implementations, the prime finite field was the main focus of many different ECC implementations for sensor networks. The choice of the field was dictated by the fact that basic arithmetic operations can be effectively optimized if pseudo-Mersenne primes are used in $F_p$. It has been stated that it is difficult to obtain an efficient ECC implementation over the binary finite field $F_{2m}$ on typical sensor platforms (Gura et al. 2004; Wang et al. 2006; Liu and Ping 2008). Binary fields have not been favored because polynomial arithmetic (multiplication, in particular) is insufficiently supported by current CPUs.

When comparing binary field arithmetic with a prime field on sensor nodes, the following misunderstandings about the implementation of ECC over $F_{2m}$ on sensor nodes are noted:

*Inefficient field multiplication*: The field multiplication in the elliptic curve operation in $F_{2m}$ is regarded as being less efficient than that in $F_p$ on low-powered and small word-sized devices, since it requires partial XOR multiplications, which are not efficiently supported by current microprocessors, at the instructional level.

*Heavy memory requirement for ECDSA*: Elliptic Curve Digital Signature Algorithms (ECDSA) implementations over $F_{2m}$ require not only field arithmetic over $F_{2m}$ but also field arithmetic over $F_p$ for generating and verifying digital signatures. Thus, it may be thought that the code size of ECDSA over $F_{2m}$ is larger than over $F_p$. Actually, most of the existing work on $F_{2m}$ only implements the Elliptic Curve Diffie Hellman (ECDH) protocol on sensor nodes. However, the code size of optimized implementations of ECDSA over $F_{2m}$ is comparable to $F_p$.

The security of ECC systems depends on several security parameters discussed in Section 9.4. Gura et al. (2004) implemented RSA and elliptic curves on 8-bit microcontrollers (Atmega128 and CC1010) and demonstrated the superiority of ECC over RSA. The implementation of the standardized RSA and ECC operations were implemented over NIST/ Standards for Efficient Cryptography Group (SECG) $F_p$ elliptic curves (NIST 1999). The RSA operations were accelerated using techniques such as Montgomery multiplication, Chinese Remainder Theorem (CRT), and Optimized Squaring. In addition, optimizations were also performed for point multiplication on elliptic curves over finite prime fields. They used projective coordinates, nonadjacent forms, and other curve-specific optimizations to improve the performance of ECC. Additionally, a hybrid multiplication algorithm was also presented that exploited the

### Table 9.1. Timings for RSA Operations on Atmega128

| | Algorithm | Time (seconds) | ROM bytes (B) | RAM bytes (B) |
|---|---|---|---|---|
| 1. | RSA-1024 public key $e = 2^{16} + 1$ | 0.43 | 1073 | 542 |
| 2. | RSA-1024 private key with CRT | 10.99 | 6292 | 930 |
| 3. | RSA-2048 public key $e = 2^{16} + 1$ | 1.94 | 2854 | 1332 |
| 4. | RSA-2048 private key with CRT | 83.26 | 7736 | 1853 |
| 5. | Modular exponentiation 512 | 5.37 | 1071 | 328 |

advantages of operand and product scanning multiplication algorithms to reduce the number of memory accesses. Timings for different RSA and ECC operations on Atmega128 clocked at 7.3728 MHz are shown in Tables 9.1 and 9.2.

The results show that the squaring of large integers and modular multiplication are one of the crucial operations for RSA and ECC. It was noted that the point multiplication operation is one of the fundamental operations in ECC, and multiprecision multiplication and squaring of large integers takes 77% of the execution time of the whole point multiplication operation on an Atmega 128-bit processor. In addition, it was observed that large integer multiplication not only involves arithmetic operations but also a significant amount of data transport to and from the memory. To optimize such operations, the authors proposed a hybrid multiplication method that combines the advantages of row-wise and column-wise multiplication.

Comparing Tables 9.1 and 9.2, point multiplication on the standard secp160r1 curve is faster than the RSA 1024 private key operation and has comparable performance to the RSA 1024 public key operation. With the increase in the security levels, we clearly see that ECC operations are much faster than RSA. Point multiplication on a secp224r1 curve takes only 2.19 seconds when compared to 83.26 seconds for the RSA-2048 private key operation, which offers the same level of security. As seen in Tables 9.1 and 9.2, memory requirements for a full RSA implementation with the Chinese Reminder Theorem are significantly higher than its counterpart in ECC. In the same environment (Atmega128 clocked at 7.3728 MHz), later implementation presented results that showed an expected time of 0.76 seconds for computing a point multiplication when compared to 0.81 seconds as achieved by this scheme (Uhsadel et al. 2007).

### Table 9.2. Timings for ECC Operations on Atmega128

| | Algorithm | Time (seconds) | ROM bytes (B) | RAM bytes (B) |
|---|---|---|---|---|
| 1. | ECC secp160r1 | 0.81 | 3682 | 282 |
| 2. | ECC secp192r1 | 1.24 | 3979 | 336 |
| 3. | ECC secp224r1 | 2.19 | 4812 | 422 |

The Non-Adjacent Forms (NAF) method was exploited for recoding the positive integer $k$ in point multiplication in order to reduce point additions. Shantz (2001) presented an efficient technique to calculate a modular division. The idea is to compute $y/x$ in one operation, instead of computing $1/x$ first and then multiplying it with $y$. This scheme has eliminated one multiplication in modular division operations. The new algorithm can be applied in both $F_p$ and $F_{2^m}$ fields.

In addition to prime fields (Section 3.3) and binary fields (Section 3.4), ECC systems can be evaluated over extension fields. Extension fields can be derived from the polynomial basis representation of binary fields. For a finite field $F_q$ where $q = p^m$ (Section 3.2.3), if $m \geq 2$, then the field $F$ is called an extension field. Woodbury et al. (2000) introduced another ECC system over an optimal extension field, where $p$ is chosen of the form $2n \pm c$ ($n$ and $c$ are arbitrary positive rational integers). The general idea with the optimal extension fields $F_{p^m}$ is to select $p$ (chosen as a small pseudo-Mersenne prime), $m$, and the reduction polynomial $f(z)$ to more closely match the underlying hardware characteristics. Usually, $p$ fits in a single computer word, and such an approach simplifies the handling of carries in arithmetic operations. Although optimal extension fields have some advantages over prime and binary fields, standardization of such fields is still not actively pursued as compared to the prime and binary fields.

ECC systems mainly consist of elliptic curve exponentiations. Generally, elliptic curve exponentiations are influenced by the coordinate—exponentiation for a fixed point and a random point. In the case of the coordinate, elliptic curve exponentiation can be computed by repeated addition and doubling operations. Furthermore, a suitable and optimized algorithm can be used to realize the repeated number of addition operations. However, realizing the doubling operations (in the case of the exponentiation for a random point) with a suitable algorithm is not possible. Coordinates on elliptic curves can be defined that provide different addition formulas. Miyaji et al. (1997) investigated the efficiency of addition formula in a Jacobian coordinate and concluded that it offers a slower addition but a faster doubling operation. In the case of exponentiation for a fixed point, a precomputation method that computes an elliptic curve exponentiation by repeating the addition operation, not the doubling operation, was initially proposed (Brickell et al. 1993). Later, Lim and Lee (1994) showed that the algorithm described in Brickell et al. (1993) can be accelerated at the cost of some extra point-doubling operations (Lim and Lee 1994). In the case of the exponentiation for a random point, a window method is used to realize the addition and subtraction operations (Koyama and Tsuruoka 1993). In this method, an interval between two windows can determine the computation amount, i.e., the longer the interval, the lesser the computation amount.

The practicality of ECC on sensor networks was first demonstrated with the introduction of the *EccM* library (Malan et al. 2004). *EccM* provided ECC over binary fields with 163-bit key sizes. To achieve better performance, it used a $y^2 + xy = x^3 + x^2 + 1$ Koblitz curve and a fixed base point. Although *EccM* showed the applicability of PKC on sensor nodes, the time taken to run ECC primitives was too high for practical applications. For instance, the time taken to generate a private and public key pair was equal to 34.2 seconds averaged over 100 trials. Furthermore, when this library was released, it was unclear if we would require hardware accelerators to improve the overall performance of the ECC applications. Later, it was shown that using projective or mixed coordinates rather than affine coordinates could avoid expensive inversions. In addition, ECC primitives were accelerated using efficient point multiplication algorithms proposed in Wang et al. (2006) and Liu and Ping (2008).

ECC implementation proposed in Wang et al. (2006) chose a prime field as the base finite field, and the experiments were carried out on a 16-bit TelosB platform involving an MSP430 microprocessor used in Tmote Sky nodes. The suite of integer arithmetic operations implemented in this operation included addition, subtraction, shift, multiplication, division, and modular reduction. The entire implementation scheme was based on an SECG-recommended 160-bit elliptic curve secp160r1 (Certicom 2000). A hybrid multiplication scheme proposed in Gura et al. (2004) was used in realizing the multiplication operation and implemented in assembly language. Random point multiplication was computed in 3.51 seconds, and fixed point multiplication was computed in 3.13 seconds. Program space utilization was at 17.9 KB, and data storage was at 1.6 KB.

In addition to the arithmetic operations, the ECDSA was also implemented. The time taken to generate ECDSA was 3.35 seconds, and the verification of the digital signature was at 6.78 seconds.

## 9.3  Design Principles

In general, any implementation on a sensor network platform should follow basic design principles that take into account portability, performance, security, and versatility.

*Portability:* Any library should ideally be compatible with a broad range of existing sensor platforms and can be easily ported to other 8-, 16-, and 32-bit constrained devices. Most of the procedures can use any high-level language, such as standard C, which favors speed and permits re-use of the code on numerous other WSN platforms. Even though some functions use inline assembly, the code-development process should be as portable as possible. Assembler routines should be generated automatically by a special utility program from user-defined macros. As a result, appropriate assembler code can be quickly developed for new platforms and processors that are not yet supported.

*Performance:* Any library should be optimized for speed and energy efficiency. Memory usage is a secondary concern, as optimizing for code size lowers the functionality and portability of the library. The memory footprint of a library can be reduced after achieving satisfactory performance by deleting unnecessary functions. To speed up the execution of particularly time-critical arithmetic routines, any high-level language code, such as the standard C code, can be replaced with assembly language specific for each platform. High performance can also be achieved through many different optimizations for elliptic curve operations.

*Security:* High levels of security in ECC systems can only be achieved through careful selection of algorithms and domain parameters. Section 9.4 provides details about

ECC security parameters. Any library should support elliptic curve parameters recommended by NIST and SECG.

*Versatility*: A large set of available library functions should provide many possibilities for writing ECC-based programs. The parameters should be flexibly configured, and the code easily integrated into WSN applications to provide the building blocks for standard ECC schemes (e.g., ECDSA, ECDH) and more advanced IBC protocols.

## 9.4 ECC Security Parameters

In general, the overall security of the system is dependent on multiple ECC parameters (Paar 1999; Hankerson et al. 2000). Although key sizes have been an important factor, there are other parameters that influence the security of the system—for instance, the elliptic curve selected for cryptography, an appropriate finite field, and the right base point. In addition, the representation used for the field elements, order of the finite field, curve coefficients in the finite field that define the equation of the elliptic curve, two field elements in finite field that define a finite point, the order of the base point, and the finite field play crucial roles in elliptic curve cryptographic systems. The number of elements in the finite field ($\#E(K)$) (Section 3.5.2) is an important factor to avoid Pollard's $\rho$ attack and Pohlig-Hellman attack on ECDLP (Pollard 1978; Pohlig and Hellman 1978). To avoid these attacks, $\#E(K)$ should be divisible by a sufficiently large prime $p$. Additionally, choosing the proper elliptic curve (in cases in which the finite field is fixed) helps improve the security of the elliptic curve cryptographic system. For instance, the maximum resistance to Pollard's $\rho$ attack can be achieved by selecting elliptic curve $E$ such that $\#E(K)$ is prime or near prime. Furthermore, $\#E(K)$ should not be equal to the order of the finite field to avoid any kind of isomorphic attacks. Isomorphic attacks reduce ECDLP to a Discrete Log Problem in groups for which subexponential time algorithms are known. It is presumed that for the special classes of elliptic curves, isomorphic attacks are faster than Pollard's $\rho$ attack. However, selecting elliptic curves at random can help prevent such attacks, provided the condition in which $\#E(K)$ is divisible by a large prime $p$ is still met.

## 9.5 The Platform

An adequate understanding of hardware limitations and capabilities is crucial to implement security. Generally, the level of security achieved is directly proportional to the processing power required. However, processing power together with memory is usually very limited on sensor devices. The processing time needed for cryptographic calculations should be short due to the short battery lifetime. When comparing security with usability, a sensor network would fail if it offers a high level of security but at the same time depletes all of the node's energy in the matter of few hours. Hence, WSN nodes require a lightweight cryptosystem.

WSN devices have to meet requirements that are usually specific to a given application, but we can still distinguish several common features. Nodes should be small, cheap, energy efficient, equipped with the right set of sensors, and have enough memory and computational power to perform given tasks. Wireless communication within the network should be highly reliable even in the face of interference and link failures. The WSN design space is very large, and hardware platforms are supposed to support more than one type of application. Due to the large variety in mote design and parameters (memory size, microcontroller type, and speed), it is hard to develop a universal security solution that would fit all hardware architectures. For cryptography implementations, available memory and specific CPU features are the most important parameters. The security solutions that target WSNs should be optimized and work with the most constrained devices (typically low-end 8-bit platforms).

From many available architectures, we can choose several devices that are the most popular platforms at the moment. Each of these motes might be regarded as a representative of a certain class of sensor devices. The MICA2DOT represents nodes with the most limited resources available. It is one of the smallest commercially available motes and is powered by a 3V coin cell. MICAz is a widely recognized 8-bit sensor platform used in many different WSN research projects. Tmote Sky is a slightly more powerful device, with 16-bit processing and a broad range of sensors embedded on the main Printed Circuit Board (PCB). In addition, Imote2 represents the most powerful group of 32-bit sensor platforms that have a lot more capabilities than previously mentioned nodes.

Cryptographic primitives are based on arithmetic operations, and their efficiency depends mainly on the hardware support of a given microprocessor. In cryptography, we have to deal with large numbers that can have more than 1000 bits. Arithmetic on such large integers is a time-consuming task, especially on 8-bit CPUs. To optimize those operations, we need to take advantage of all the specific features of a given device.

Typical devices have 8- or 16-bit architectures and clock speeds around 4–8 MHz. The data memory available can be as low as 4 KB, and program storage might be limited to only 48 KB. These constraints make it difficult to implement some of the more complex security primitives (as required in PKC).

### 9.5.1 MICAz/MICA2 Mote

The MICAz mote sensor node is equipped with an ATmega128 8-bit processor clocked at 7.3728 MHz. The program code is loaded from an 128-KB EEPROM chip, and runtime memory is stored in a 4-KB RAM chip. The ATmega128 processor is a typical RISC architecture with 32 registers, but six of them are special pointer registers. Since at least one register is needed to store temporary results or data loaded from memory, 25 registers are generally available for arithmetic. The instruction set is also reduced, as only 1-bit shift/rotate instructions are natively supported. Bitwise shifts by arbitrary amounts can then be implemented with combinations of shift/rotate instructions and other instructions. The processor pipeline has two stages, and memory instructions always cause pipeline stalls. Arithmetic instructions with register operands cost 1 cycle, and memory instructions or memory addressing cost 2 processing cycles.

### 9.5.2 16-bit MSP430 Microcontroller

The 16-bit MSP430F1611 produced by Texas Instruments differs in many ways from the Atmel chip. The MSP430 has a more traditional architecture that uses mainly memory-to-memory operations rather than the more classic RISC load-store approach of the ATmega128. Its instruction set is limited to 27 instructions, but the variety of 7 different addressing modes offers a lot of flexibility in data manipulation. This orthogonal architecture allows every instruction to be used with every addressing mode. An interesting feature of this CPU is the existence of an external hardware multiplier that performs fast multiplication of 8- and 16-bit integers. The MSP430 provides 12 general-purpose registers. The remaining 4 registers, r0–r3, are used for Program Counter, Stack Pointer, Status Register, and Constant Generator, respectively. Registers r4–r15 are general purpose and available for use at all times. This CPU also offers 10 KB of RAM, but only 48 KB of ROM, which might be a problem in the case of large programs. A very nice feature of the MSP430 is its ultra low power consumption, which is especially important on tiny devices in distributed environments where the available energy is very limited.

### 9.5.3 Marvell PXA271

The 32-bit Marvell PXA271 processor is built around the ARM core and differs completely from the two previously described platforms. It is a much more advanced and powerful CPU, with superior capabilities. In addition, the memory resources on this processor are a few orders of magnitude higher than on the other two devices (32 MB of ROM and 256 KB of RAM). This range of CPUs is commonly used in smartphones (e.g., the iPhone) and PDA class devices and opens new possibilities when used in the WSN environment. The ARM processor has a standard RISC load-store architecture, with several innovative features, including effectively free shifting of operands at no extra cost (using a built-in barrel shifter) and conditional execution of instructions. It has 16 registers, although 3 of these are reserved for special purposes—r13 is a stack pointer, r14 holds a function return address, and r15 is the program counter. If needed, the content of r14 can be pushed onto the stack, and an extra register is then available until the end of the called function. The PXA271 offers full 32-bit processing, and its clock rate can be changed dynamically by adjusting the input voltage. This interesting feature permits a tradeoff between energy consumption and processing power during the CPU operation.

Table 9.3 brings together many important parameters of popular WSN platforms. This table shows that most of the popular sensor platforms have limited computational and storage capacity. Hence, implementing security on such resource-constrained

Table 9.3. Comparison of Popular Wireless Sensor Platforms

| | Processing | CPU | RAM | Flash | Frequency | Receive power | Transmit power | Data rate |
|---|---|---|---|---|---|---|---|---|
| Mica2DOT | 8-bit | Atmega128L | 4 kB | 512 kB | 433 MHz | 29 mW | 42 mW | 38.4 kb/s |
| Mica2 | 8-bit | Atmega128L | 4 kB | 512 kB | 433 MHz | 29 mW | 42 mW | 38.4 kb/s |
| Tmote Sky | 16-bit | MSP430 | 10 kB | 1 MB | 2.4 GHz | 38 mW | 35 mW | 250 kb/s |
| Imote2 | 32-bit | PXA271 | 256 kB | 32 MB | 2.4 GHz | 38 mW | 35 mW | 250 kb/s |

platforms would require optimization. Among the platforms presented in the above table, the most popular platforms are MICA2DOT, MICA2, and MICAz, with an 8-bit ATmega128 processor by Atmel Corporation.

## 9.6 Related Work

Recent technological advances have made it possible to develop WSN consisting of a large number of low-cost, low-power, and multifunctional sensor nodes that communicate over short distances through wireless links. Such sensor networks are ideal candidates for a wide range of civilian and military applications (Section 1.1.1). The desirable features of WSN have attracted many researchers to develop protocols and algorithms that can fulfill the requirements of these applications. For instance, security services such as authentication, confidentiality, and key management are critical to communication in WSN as well as the security of sensor network applications. In traditional networks such as the Internet, PKC has been the enabling technology underlying many security services and protocols (e.g., Transport Layer Security [TLS] and Internet Protocol security [IPsec]). In some instances, PKC has been used to bootstrap symmetric session keys and authenticate messages to multiple receivers. However, in WSN, PKC has not been widely adopted due to resource constraints on sensor platforms, particularly due to the limited and depleting battery power.

In recent years, there has been extensive research aimed at developing techniques that aim towards circumventing PKC operations in sensor network applications. For example, there has been a substantial amount of work done on random key predistribution for pairwise key establishment (Section 6.3.4) and broadcast authentication (Section 5.1). However, these alternative approaches do not offer the same degree of security or functionality as PKC. For instance, none of the random key predistribution schemes can guarantee key establishment between any two nodes and tolerate arbitrary node compromises at the same time. As another example, broadcast authentication schemes, which are all based on µTESLA (Section 5.2.1.2), require loose time synchronization, which itself is a challenging task to achieve in WSN. In contrast, PKC can address all the abovementioned issues. Pairwise key establishment can be achieved using, for example, the DH key exchange protocol, without suffering from the node compromise problem. Similarly, broadcast authentication can be provided with, for example, the ECDSA digital signature scheme without requiring time synchronization. Hence, it is desirable to explore the application of PKC on resource-constrained sensor platforms (Section 9.1).

Despite the recent progress of ECC implementations on sensor platforms, all the previous attempts have limitations. In particular, all these attempts were developed as independent packages/applications, without seriously considering the resource demands of sensor network applications. As a result, developers may find it difficult, and sometimes impossible, to integrate an ECC implementation with the sensor network applications, although the ECC implementation may be viable on its own. For example, an ECC implementation may require so much RAM that it would be impossible to fit both the sensor network application and the ECC implementation on the same node.

Moreover, various optimization techniques are available to speed up the ECC operations. Such optimizations, however, typically will increase the ROM and RAM consumptions, although they may reduce the execution time and energy consumption. It is not clear what optimizations should be used and how they should be combined to achieve the best trade-off among the following: security protection, computation overheads, and storage requirements. Additional research is necessary to clarify these issues and facilitate the adoption of ECC-based PKC in WSN.

It is certainly possible to have dedicated PKC hardware included on sensor platforms. However, given that there is no PKC hardware that is currently available on current sensor platforms, it is a viable choice to explore software approaches for PKC support on these platforms.

In this chapter, we review optimized techniques for arithmetic in binary and prime fields—for squaring, multiplication, and modular reduction on popular sensor platforms such MICAz or MICA2 mote.

We look at the performance issues of ECC-based systems and compare the cost associated with the multiplication of a random point on an elliptic curve by a random integer. In addition, we review implementations of modular reduction algorithms and field multiplication that concentrate on the reduction of memory access and increased computational speed.

## 9.6.1 TinyECC

ECC that was optimized for resource-constraint network platforms such as sensor networks was presented in TinyECC (Liu and Ping 2008). ECC operations include authentication, encryption algorithms, and key exchange mechanisms. It uses several optimization techniques, such as a projective coordinate system, curve-specific optimization, a sliding window method, and a hybrid multiplication method.

*Projective coordinate system*: In many of the 8-bit microprocessors such as ATmega128, the inversion operation is significantly more expensive than the multiplication operation. Hence, such processors do not support the division instruction. As a result, it is more efficient to implement elliptic curve operations in projective coordinates than affine coordinates. In this scheme, weighted projective representation, such as Jacobian coordinates, are used to expedite point addition, point doubling, and scalar point multiplication.

*Curve-specific optimization*: For all NIST and most SECG curves, pseudo-Mersenne primes were chosen as the underlying field primes to allow optimized modular reduction. This optimized modular reduction algorithm was implemented in order to expedite modular squaring and multiplication operations.

*Sliding window method:* This algorithm was used to expedite scalar point multiplication by reducing the total number of point additions. However, one of the disadvantages of

**Table 9.4. TinyECC Performance Analysis and Memory Utilization on Different Sensor Platforms**

|  | Platform | Time (seconds) | ROM kilobytes (kB) | RAM kilobytes (kB) |
|---|---|---|---|---|
| 1. | ATmega128 at 7.3728 MHz | 2.1 | 16 | 1.8 |
| 2. | TelosB at 4 MHz | 4.2 | 11.4 | 1.8 |
| 3. | Imote2 at 13 MHz | 0.4 | 12.9 | 2.1 |

this method is the requirement of precomputation of some points that would require extra memory.

*Hybrid multiplication method:* This method combines features of column-wise and row-wise multiplication (discussed in the later sections) and leverages on the processor's architecture as it has many available general-purpose registers.

TinyECC is intended for sensor platforms running TinyOS. It is implemented in nesC, with additional platform-specific optimizations in inline assembly language for popular sensor platforms. So far it has been tested on MICAz, TelosB, Tmote Sky, and Imote2. TinyECC 1.0 supports SECG-recommended 128-, 160-, and 192-bit elliptic curve domain parameters. In particular, the following curves are used: sec128r1, secp128r2, secp160k1, secp160r1, secp160r2, secp192k1, and secp192r1 (Certicom 2000). Table 9.4 shows the performance analysis point multiplication and memory utilization on three different sensor node platforms for a finite prime field.

## 9.6.2 NanoECC

Efficient implementation of ECC primitives in sensor networks is a difficult task. Small memory size, limited CPU capabilities, and scarce battery resources have created a difficult environment for applying ECC. Most of the existing implementations focus on optimizing ECC for a particular hardware platform and do not focus on the portability of the code. Many available libraries use fixed parameters, which does not permit the changing of elliptic curves, domain parameters, or security levels. The NanoECC library offers efficient ECC primitives that can be used to build complete security protocols (Szczechowiak et al. 2008). This library can be integrated into existing sensor network applications to provide PKC-based security solutions. In addition, it is optimized for different sensor network platforms and can be flexibly configured to find the best set of parameters for a given application type. It is compatible with MICA2, MICA2DOT, MICAz, TelosB, and Tmote Sky motes but can also be easily ported to other hardware platforms.

The library is based on the four design principles discussed in Section 9.3. It provides ECC operations over both the prime field $F_p$ and the binary field $F_{2m}$. The library

**Table 9.5. Fixed Point and Random Point Multiplication Using NanoECC Library**

| Field | Fixed point multiplication | | | | Random point multiplication | | | |
|---|---|---|---|---|---|---|---|---|
| | $F_{2m}$, $m = 163$ | | $F_p$, $p = 160$ | | $F_{2m}$, $m = 163$ | | $F_p$, $p = 160$ | |
| Language | C | C+asm | C | C+asm | C | C+asm | C | C+asm |
| Timings (seconds) | 2.97 | 0.66 | 1.20 | 0.60 | 2.16 | 0.87 | 2.59 | 0.99 |
| ROM (kB) | 32.4 | 38.7 | 57.8 | 40.1 | 32.4 | 38.7 | 57.8 | 40.1 |
| RAM (kB) | 1.7 | 2.1 | 1.8 | 1.9 | 1.7 | 2.1 | 1.8 | 1.9 |

can also be used in conjunction with different embedded operating systems. Results achieved by NanoECC using both types of finite fields—prime and binary—show, for the first time, that in some cases in this constrained environment, ECC operations over the binary polynomial field $F_{2m}$ outperform those in $F_p$. Moreover, it was shown that the timings for the binary field case would be even faster if a binary polynomial multiplication instruction was available on the architectures considered.

The core of NanoECC is based on the Multiprecision Integer and Rational Arithmetic C/C++ Library (MIRACL), which provides all the necessary tools to perform operations on elliptic curves. MIRACL supports ECC over both standard finite fields, $F_p$ and $F_{2m}$. It also handles all of the operations on big integers and large polynomials. MIRACL was originally designed for desktop class computers but can also be optimized to work with constrained 8- and 16-bit platforms. On a PC with large memory space, all the memory for big integers is allocated dynamically from the heap. In an embedded environment, the memory resources are very limited, and dynamic memory allocation is often not available. MIRACL had to be optimized to allocate all the memory exclusively from the stack. One downside of this is that the maximum size of big variables has to be set at compile time, when the library is being created. Nevertheless, static memory allocation allows maximum use and re-use of memory, and avoids fragmentation of precious RAM. The MIRACL library was extensively optimized to provide the best ECC performance on constrained WSN motes. Table 9.5 summarizes fixed point and random point multiplication using the NanoECC library.

Recently, TinyECCK presented an implementation of ECC over binary curves (Seo et al. 2008), which takes into account the platform characteristics to optimize finite field arithmetic and obtained a point multiplication in 1.14 seconds.

## 9.7 Implementation of Identity-Based Cryptography (IBC) in WSN

ECC uses smaller PKC parameters than RSA, making it more applicable to WSN (discussed in previous sections). Although ECC has been used as the underlying cryptography to implement many security protocols, its use in IBC protocols helps provide practical secure solutions to resource-constraint environments such as WSN. As mentioned in Chapter 4, IBC is a public key technique that is based on the identities of entities. In the WSN environment, the sensor node's identity can be used as the public key. Hence, there is no need for a certificate to bind a sensor node's identity to its public key. Therefore, IBC provides practical public key cryptographic schemes (authentication, encryption, or session key setup) without the use of a complex public key infrastructure. In many ways, an identity-based scheme is an appropriate solution for sensor networks. There is no need to maintain a public key directory, as the public keys can be derived from node identities that are widely known in the network. IBC provides a scalable security mechanism in which the number of keys is kept to a minimum. Nodes generate a public key for a given node only when they want to communicate with it for the first time. After agreeing upon a shared session key, nodes can use symmetric key mechanisms to encrypt subsequent messages and to communicate in a secure manner.

IBC allows every node to send secure messages to all the other nodes from the beginning of the network operation. No prior interaction between the nodes is needed. Exchange of the information does not require any service or assistance from a third party. However, identity-based systems assume the existence of a trusted key-generation center, which issues private keys corresponding to user identities. This authority can use its master key to decrypt user messages. It can also impersonate anyone in the network. This feature introduces the key escrow problem, wherein the security of the whole system depends on the public key generator security. In many cases, a single unconditionally trusted entity in the network simply does not exist. Fortunately, in sensor networks the original network deployer can be considered as a trusted entity that can act as the public key generator. It can generate a unique secret key based on each node's identity and pre-load this information to the node's memory before the deployment phase. At this stage, a secure channel clearly exists which allows careful configuration of the network. The application of IBC to WSN is presented in Figure 9.1.

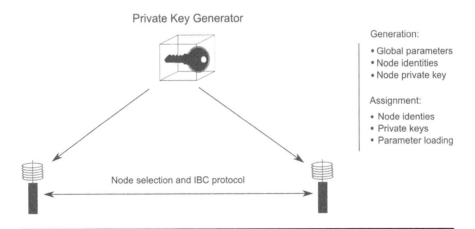

**Figure 9.1.** Identity-Based Cryptography in WSN

Security schemes that are based on IBC allow the easy addition of new nodes to the network. There is no need to replace or add new keys to existing devices. New sensor nodes only have to be programmed with the domain parameters and a private key by the public key generator before deployment. The communication overhead in establishing session keys is minimal in IBC schemes. This makes the IBC approach much more suitable for low energy WSNs than traditional public key infrastructure schemes. The elimination of digital certificates lowers the energy consumption and makes the system more practical, especially in WSNs that are deployed in remote areas. In the case of sensor networks, IBC offers better security than other methods that are not based on PKC. The whole system also increases resistance against a node-capture attack. Subverting one of the nodes does not reveal anything about the communication between other pairs of nodes. It only allows decryption of the messages received from other nodes. Network access control is also provided, as only the public key generator issues identities and pre-loads sensors with valid private keys. An active attacker can encrypt messages to given identities but cannot decrypt any message without a proper private key. IBC has clear advantages over traditional public key systems but also has some inherent problems. One of them is key revocation. In traditional PKC systems, compromised keys are replaced with a new private/public key pair. In IBC systems, key revocation requires that users have to change their identity information that corresponds to given private keys. This might be especially problematic in cases in which identities are chosen as the nodes' unique physical addresses (e.g., transceiver serial numbers). One solution to the problem might be to use network addresses (e.g., IPv6 addresses) to identify nodes in the network. An alternative solution would be to combine the date with the identity information to generate new private keys, when necessary. The problem of key revocation highlights the importance of proper management of node identities in IBC systems.

Despite some minor problems, IBC has many advantages when compared with other security schemes. Table 9.6 summarizes the main benefits that arise from using IBC with secure WSN.

Implementation of IBC protocols is more complicated than traditional public key cryptographic schemes (e.g., RSA). IBC protocol implementation is based on the efficient implementation of modules, such as the following: finite field arithmetic, big number and modular arithmetic, ECC arithmetic, and cryptographic pairing for implementing pairing-based cryptographic libraries.

*Finite field arithmetic*: Elliptic curves used in cryptography are defined over finite fields. Therefore, efficient arithmetic operations in the underlying field are crucial to the overall performance of the system. Much effort is devoted in expediting basic arithmetic routines, as they are frequently used in higher level operations. However, with varying sensor network platforms, platform-dependent hand-coded assembly routines are used to increase the speed of basic arithmetic routing. Although this approach deviates from the basic design principles stated in Section 9.3, it helps improve the efficiency of the overall system. Hence, assembly routines that are platform dependent can be part of the cross-layer design approach in WSN.

Table 9.6.  IBC in Comparison with PKC and IBC

| | Key distribution | Number of keys | Key directory | Digital certificates | Forward encryption | Nonrepudiation |
|---|---|---|---|---|---|---|
| Symmetric key cryptography | Problematic | $O(n^2)$ | At each node | No | No | No |
| Random key predistribution | Simple | $O(n)$ | At each node | No | No | No |
| PKC | Complex | $O(n)$ | At each node and key center | Yes | No | Yes |
| IBC | Simple | N | No | No | Yes | Yes |

*Big number and modular arithmetic*: In cryptography, arithmetic operations are performed on numbers that are hundreds of bits long. Such numbers are not supported natively in many computing languages and must be implemented externally. Appropriate handling of big numbers and modular arithmetic is a key to efficient cryptographic operations.

*Elliptic curve arithmetic*: Cryptographic mechanisms based on elliptic curves depend on arithmetic involving the points of the curve. Elliptic curve operations rely on arithmetic in the finite field, but the performance of point addition and doubling is also important. The efficiency of curve arithmetic depends on the type of the curve, the size of the underlying field, the point representation system, and the implementation of the algorithms.

*ECC primitives*: One of the main cryptographic primitives in ECC is scalar point multiplication. This operation is the basis for many ECC schemes, such as ECDSA, Elliptic Curve Diffie-Hellman, and other elliptic curve encryption schemes. The performance of point multiplication is crucial in ECC, as it dominates the execution time of the above schemes. Point multiplication is an important primitive, not only in standard ECC, but also in IBC. Efficient implementation of point multiplication is difficult due to the variety of existing algorithms and different types of elliptic curves.

*Cryptographic pairings*: The calculation of cryptographic pairings is currently the most computationally intensive operation in IBC schemes. Software implementations of pairings are considered complex and computationally expensive for embedded devices, and the application of Pairing-Based Cryptography (PBC) on sensor networks is an emerging field in WSN security. For instance, for an 80-bit security level, PBC works with 1024-bit numbers, as opposed to the conventional 160-bit numbers used in ECC.

## 9.8 Algorithms for Finite Field Arithmetic

Finite field arithmetic forms the fundamental basis for building identity-based cryptographic protocols (Szczechowiak 2010). In general, finite fields portray special properties that could be utilized for cryptography and especially ECC, as elliptic curves used in cryptography are defined over finite fields. In addition, curve operations such as addition, subtraction, multiplication, squaring, and inversion are performed on the underlying field elements. Due to the repeated use of these operations, the algorithms used for implementing field arithmetic plays a crucial role in increasing the efficiency of the system. Additionally, the type of field (prime fields [Section 3.3]; binary fields [Section 3.4]; or optimal extension fields) and its size also influence the overall performance of the system. Among the arithmetic operations, multiprecision multiplication of field elements is the most time-critical operation, and the efficiency of a low-end processor (8-bit Atmega 128 CPU, 16-bit MSP430 microcontroller, and 32-bit ARM processor) is largely dependent on its effective implementation.

## 9.8.1  Prime Field Arithmetic on Resource-Constraint Processors

Typical prime field arithmetic operations include addition, subtraction, squaring, multiplication, and reduction operations. Field inversion is also required for some operations, but in many cases it can be replaced with a larger number of cheaper operations. For example, using Montgomery's method, the modular reduction can be carried out without using division, and, hence, it is the modular multiplication and squaring that are significant (Montgomery 1985). Multiprecision multiplication is the time-critical requirement in a great majority of number-theoretic–based methods for PKC. It is required for the RSA algorithm, for methods based on elliptic curves, and also for new mechanisms that are based on cryptographic pairings. Elliptic curve systems are usually defined over finite fields $F_p$ of large characteristic (Section 3.2.3), where $p$ has at least 160 bits (for security reasons). Since the modulus is of a fixed size in bits, the code for addition, multiplication, squaring, and modular reduction can be written in assembly language. The loops can be completely unrolled to achieve maximum performance at the cost of some additional storage.

The implementation of prime field arithmetic requires that each field element is represented using multiple machine words. In the case of $F_p$, where $p$ is a large prime, the field element $a$ can be implemented as a series of $W$-bit unsigned integers $0 < a_i < 2^W$, where $W$ is the word size on the target machine (e.g., $W = 8, 16, 32$) and $a = \sum_{i=0}^{t-1} a_i 2^{W_i}$. The number of required words $t$ to hold a field element can be calculated from $t = \left\lceil \dfrac{\log_2 p}{W} \right\rceil$. The binary representation of a field element $a$ can be stored in an array $A = (A[t\text{-}1], \ldots, A[2], A[1], A[0])$, where the rightmost bit of $A[0]$ is the Least Significant Bit (LSB).

### 9.8.1.1  Modular Addition and Subtraction

Multiprecision integer addition and subtraction is a straightforward operation on standard general-purpose processors. It requires $t$ additions (or subtractions) of word-size integers. On processors that have the add-with-carry (and subtract-with-carry) instruction, there is no need for an explicit check for carry. Arithmetic in the field $F_p$ requires modular additions $c = (a + b) \bmod p$ and subtractions $c = (a − b) \bmod p$. These operations require an additional step, which performs reduction modulo $p$. In the case of modular subtraction, after calculating $t$ word-size subtractions, the borrow bit is tested. If the borrow is equal to 1 then the modulus $p$ is added to the result $c$. Similarly, in modular addition, the carry bit decides if a reduction is needed. When the carry bit is equal to 1, then the modulus $p$ needs to be subtracted from $c$. The reduction is also needed if carry is equal to 0 and $c \geq p$.

Figure 9.2 presents the timings for modular addition and subtraction operations on 160-bit integers in $F_p$. All the timings are given in clock cycles of a given hardware

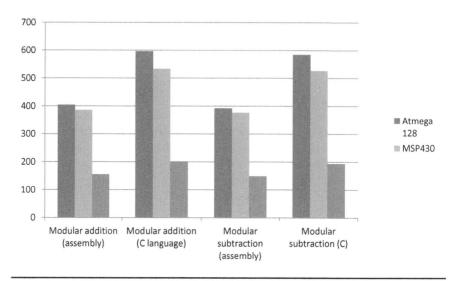

**Figure 9.2.** Timings in Instruction Cycles for Modular Addition and Subtraction of 160-Bit Integers on Standard WSN Processors

platform. They represent an average value over 100 operations because the reduction step is not always executed.

### 9.8.1.2 Multiprecision Integer Multiplication

There are two main methods to implement multiprecision integer multiplication in software. The first group uses the Karatsuba-Ofman technique, which divides the operands into smaller integers and thus reduces the number of multiplications. Unfortunately, this approach has a large overhead in addition operations and is not efficient in finite fields of practical interest. The second group of techniques is more practical and uses different variations of basic schoolbook multiplication. When doing multiplication using the schoolbook algorithm, the multiplication is divided into several partial products that are accumulated to get the final result. The partial products can be calculated in any order before they are added together. They can be arranged in rows from right-to-left or in columns by bit length. Figures 9.3 and 9.4 show a 4×4 multiplication using both methods.

The column-wise technique (also known as the product scanning method) is not as straightforward to program since the columns are of different lengths, although this is not a concern if the code is unrolled. The alternative row-wise method (also known as the operand scanning method) is much easier to implement as a short looped program. The word-by-word multiplication can be carried out using a simple pair of nested "for" loops. The row-wise algorithm requires more memory writes, but less memory reads,

| $c_7$ | $c_6$ | $c_5$ | $c_4$ | $c_3$ | $c_2$ | $c_1$ | $c_0$ |
|---|---|---|---|---|---|---|---|
|  |  |  |  | $a_3$ | $a_2$ | $a_1$ | $a_0$ |
|  |  |  |  | $b_3$ | $b_2$ | $b_1$ | $b_0$ |
|  |  |  |  |  |  | $a_0$ | $b_0$ |
|  |  |  |  |  | $a_1$ | $b_0$ |  |
|  |  |  |  |  | $a_0$ | $b_1$ |  |
|  |  |  |  | $a_2$ | $b_0$ |  |  |
|  |  |  |  | $a_1$ | $b_1$ |  |  |
|  |  |  |  | $a_0$ | $b_2$ |  |  |
|  |  |  | $a_3$ | $b_0$ |  |  |  |
|  |  |  | $a_2$ | $b_1$ |  |  |  |
|  |  |  | $a_1$ | $b_2$ |  |  |  |
|  |  |  | $a_0$ | $b_3$ |  |  |  |
|  |  | $a_3$ | $b_1$ |  |  |  |  |
|  |  | $a_2$ | $b_2$ |  |  |  |  |
|  |  | $a_1$ | $b_3$ |  |  |  |  |
|  | $a_3$ | $b_2$ |  |  |  |  |  |
|  | $a_2$ | $b_3$ |  |  |  |  |  |
| $a_3$ | $b_3$ |  |  |  |  |  |  |
| + | + | + | + | + | + | + | + |
| $c_7$ | $c_6$ | $c_5$ | $c_4$ | $c_3$ | $c_2$ | $c_1$ | $c_0$ |

Figure 9.3. Column-Wise Multiplication

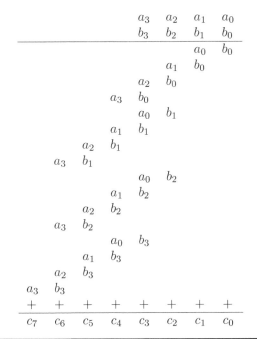

| $c_7$ | $c_6$ | $c_5$ | $c_4$ | $c_3$ | $c_2$ | $c_1$ | $c_0$ |
|---|---|---|---|---|---|---|---|
|  |  |  |  | $a_3$ | $a_2$ | $a_1$ | $a_0$ |
|  |  |  |  | $b_3$ | $b_2$ | $b_1$ | $b_0$ |
|  |  |  |  |  |  | $a_0$ | $b_0$ |
|  |  |  |  |  | $a_1$ | $b_0$ |  |
|  |  |  |  | $a_2$ | $b_0$ |  |  |
|  |  |  | $a_3$ | $b_0$ |  |  |  |
|  |  |  |  |  | $a_0$ | $b_1$ |  |
|  |  |  |  | $a_1$ | $b_1$ |  |  |
|  |  |  | $a_2$ | $b_1$ |  |  |  |
|  |  | $a_3$ | $b_1$ |  |  |  |  |
|  |  |  |  | $a_0$ | $b_2$ |  |  |
|  |  |  | $a_1$ | $b_2$ |  |  |  |
|  |  | $a_2$ | $b_2$ |  |  |  |  |
|  | $a_3$ | $b_2$ |  |  |  |  |  |
|  |  |  | $a_0$ | $b_3$ |  |  |  |
|  |  | $a_1$ | $b_3$ |  |  |  |  |
|  | $a_2$ | $b_3$ |  |  |  |  |  |
| $a_3$ | $b_3$ |  |  |  |  |  |  |
| + | + | + | + | + | + | + | + |
| $c_7$ | $c_6$ | $c_5$ | $c_4$ | $c_3$ | $c_2$ | $c_1$ | $c_0$ |

Figure 9.4. Row-Wise Multiplication

of the operands. Therefore, the column-wise algorithm should be preferred, as memory writes only occur at the foot of each column. A disadvantage of both methods is that they reload the same operands many times to calculate different partial products. This is obviously not an optimal approach, especially if a processor has enough registers available to store some of these values for later use.

It has recently been suggested that a hybrid method, which combines features of both techniques, might be preferable, particularly in a setting where the processor has many available general-purpose registers (Gura et al. 2004). This method was first described in the context of the ATmega-128L 8-bit processor, which has 32 registers. In the same setting, Uhsadel et al. (2007) have presented better results. The following sections present a description of an optimized method of hybrid multiplication that simplifies the whole approach. This improved technique can be applied to a wide range of embedded processors and is superior to the earlier proposals.

### 9.8.1.3 Improved Hybrid Multiplication

Most efficient implementations of multiprecision multiplication use the Comba method, which is based on the column-wise technique (Figure 9.3; Comba 1990). Each $W$-bit pair of digits is multiplied together to create one $2W$-bit partial product. This partial product is accumulated in a triple register. The third register (called a "carry-catcher") is required to catch the carries that can arise as the full column is added up. After the column addition, the least significant register of the triple register is written to memory as a part of the result, and the other two registers shift down, representing the carry to the next column. The maximum number that can arise in the third and most significant register is bounded by the number of digits in the multiprecision multiplication. This is likely to be much smaller than $2^W$-1, which is the maximum number that can be stored in a register. A smaller register can, therefore, sometimes be used here.

The hybrid method maintains the advantages of the column-wise method while exploiting any extra registers to avoid unnecessary reloads of data (Gura et al. 2004). The idea is straightforward: perform the multiplication as if the word length of the computer were actually $d.W$, and perform the $d{\times}d$ multiplication that arises in the calculation of the larger partial product using the row-wise algorithm. Figure 9.5 illustrates the hybrid method for the case $d = 2$. Four word-size integers are loaded into registers. As each large 2×2 partial product is calculated (represented by the large outer boxes), it must be accumulated into registers. This figure illustrates the accumulation of a particular partial product, and curved arrows indicate the carries. A naive implementation would require five registers in this case to store the sum. However, in the worst case this might require up to five add-with-carry instructions because, as in integer addition, a carry-out is always a possibility that must be considered. The original hybrid method (Gura et al. 2004), as well as the improved variant described in Uhsadel et al. (2007), use a rather complex method to deal with this carry propagation problem based on the fact that the product of two words, plus two words, cannot overflow a double-word-sized register since $(2^W - 1)(2^W - 1) + (2^W - 1) + (2^W - 1) = 2^{2W} - 1 < 2^{2W}$.

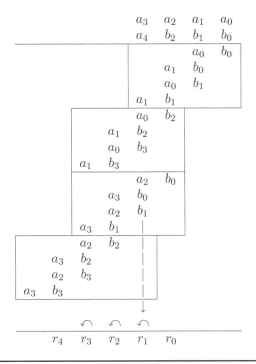

**Figure 9.5.** A Naive Implementation of the Hybrid Method

A new strategy for improving the hybrid method is by employing more registers to suppress the carry propagation. As well as using $2d$ column registers to hold the sum of each now wider column, extra $2d-1$ registers are deployed as "carry catchers." These registers are initialized to zero at the top of the column and catch any carries that may arise, without the possibility of further carry propagation. When the column is finished, these carry catchers are simply added with carry to the main set of column registers.

The case for $d = 2$ is shown in Figure 9.6. The most significant carry-catcher register performs exactly the same role as the fifth register in the native implementation. The $d{\times}d$ row-wise multiplication requires $d + 1$ registers: $d$ registers to hold the entire row, each element of which is then multiplied by another register (which stores the current multiplier) to create $d$ partial products. There are $2d$ column registers and $2d-1$ carry-catchers, for a total of $5d$ registers. For $d = 1$, this is the original Comba method, which requires five registers, just possible if the number of general-purpose registers is $r = 8$. The choice $d = 2$ requiring 10 registers is a good fit for a processor with a total of $r = 16$ registers, and $d = 4$ will take 20 registers for the case $r = 32$. The improved hybrid method can exploit even more registers if they are available on the target platform. The above figures are approximate and may require modification for a particular architecture. Not all of the $r$ registers on a given type of CPU might be available to the developer. At least three registers will be required to store the memory address of the operands and of the result.

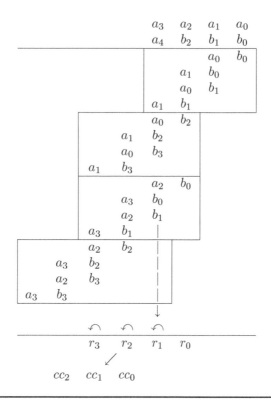

**Figure 9.6.** The Improved Hybrid Implementation

It is also sometimes helpful if a register fixed with the constant value of zero is available. The processor must support simple indexed memory load and store instructions, an integer multiply instruction that outputs the result in a pair of registers, and a simple add-with-carry instruction. The implementation of the improved hybrid method is always processor specific, as CPUs differ in small but significant ways that cannot be described in a one-size-fits-all way.

## 9.8.2 Efficient Binary Field Arithmetic for Low-End Processors

Finite fields of characteristic-two are usually the main alternative to prime fields when implementing ECC systems. Binary fields are attractive, especially for hardware implementations, since the operations in $F_{2^m}$ involve only shifts and bitwise addition modulo 2. The simplicity of basic operations is also attractive for software implementations on general-purpose processors. Prime and binary fields have their own pros and cons when it comes to efficient implementation on low-end processors. The most time-critical operation in $F_{2^m}$ is binary polynomial multiplication, which involves calculations of a few bits at a time and can be very slow without proper optimizations. On the other hand, the arithmetic in prime fields is typically more difficult to implement efficiently, due in part to the propagation of carry bits. Nonetheless, prime fields are usually preferred due to better hardware support on embedded processors.

Most of the low-end CPUs support multiplication of word-size integers in hardware. This feature significantly accelerates multiprecision multiplication in $F_p$. Arithmetic on binary fields is not natively supported in the Instruction Set of typical embedded processors. None of the CPUs used in sensor networks support multiplication in the binary field. Therefore, efficient implementation of polynomial multiplication remains a challenging task. The elements of a finite field $F_{2^m}$ can be represented by binary polynomials of degrees less than $m$, where $m$ is a prime number. For security reasons, ECC systems are defined over finite fields where $m$ is usually greater than 160 bits. In software, the coefficients of a binary polynomial may be stored in an array of $tW$-bit words. The $s = tW-m$ highest order bits in the last word of the array remain unused (always set to 0).

The addition of field elements is performed bitwise and is very fast, requiring only $t$ word operations (XORs), which are supported by all computer architectures. Addition in the binary field is analogous to multiprecision integer addition, without the carries. Other arithmetic operations (beside multiplication) include inversion, squaring, reduction, and calculation of square roots.

### 9.8.2.1 Binary Polynomial Multiplication

The simplest way to perform polynomial multiplication is to use basic schoolbook multiplication. This method for two polynomials $p(z)$ and $q(z)$ of degree $m-1$ is illustrated in Figure 9.7.

The algorithm requires $m-1$ left shifts of $p(z)$ and a modulo 2 addition of $p(z)$ to the accumulator for every non-zero coefficient of $q(z)$. This "shift-and-add" method

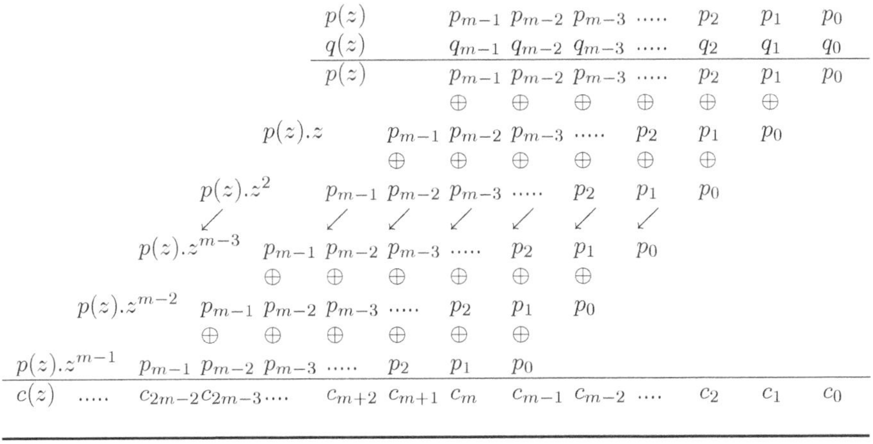

**Figure 9.7.** Binary Polynomial Multiplication Using the Basic Schoolbook Method

generates only one bit of the result at each step. Its implementation in software is usually very slow due to many vector shifts. The large number of these operations renders the implementation unfeasible, especially on small 8-bit processors where shifts can only be performed one bit at a time. The shift-and-add method is more suitable for hardware, where a vector shift can be performed in one clock cycle. Embedded processors require faster methods for field multiplication. The following sections present a description of an optimized hierarchical technique for binary polynomial multiplication. This method combines different multiplication techniques and achieves superior results on constrained WSN platforms.

### 9.8.2.2 Optimized Hierarchical Method for Polynomial Multiplication

The optimized hierarchical technique for binary polynomial multiplication consists of two stages: the first stage is the application of the Karatsuba-Ofman technique, which divides large polynomials into several smaller polynomials (Knuth 1997); the second stage performs the actual multiplication of binary polynomials using the Comb method (Lopez and Dahab 2000).

For embedded processors with small word sizes, the hierarchical multiplication technique is attractive for large fields. In the case of smaller values of $m$, the second stage of the technique can be applied directly in order to avoid unnecessary overhead. The execution time of the whole method depends on the efficiency of the Comb multiplication. This algorithm should be implemented in assembly language to provide maximum performance on a given hardware platform. The multiplication of two binary polynomials $a(z)$ and $b(z)$ of degree $m$-1 results in a polynomial of degree at most $2m$-2. In the first step, the Karatsuba-Ofman method divides large $m$-1 polynomials into several polynomials with fewer coefficients. This divide-and-conquer method can be directly adapted for binary polynomials. In the case of a division into two parts (depth $d = 2$), the polynomial multiplication $c(z) = a(z)b(z)$ can be represented as

$$a(z) = A_1(z)x^n + A_2(z) \tag{9.1}$$

$$b(z) = B_1(z)z^n + B_2(z) \tag{9.2}$$

$$c(z) = A_1 B_1 z^{2n} + ((A_1 + A_2)(B_1 + B_2) + A_2 B_2 + A_1 B_1)z^n + A_2 B_2 \tag{9.3}$$

where $n = \left\lceil \dfrac{m}{2} \right\rceil$ and $A_1, A_2, B_1, B_2$ are binary polynomials of degree less than $n$.

For $d = 2$, the above technique substitutes one full multiplication with three half-size multiplications. One multiplication is saved at the cost of several extra additions that are very fast in binary fields. The most time-consuming part of Eq. (9.3) is the computation of the subproducts $A_1 B_1, A_2 B_2$ and $(A_1 + A_2)(B_1 + B_2)$. In the second stage of the hierarchical multiplication, these products are calculated using the optimized Comb method with a window width of $w$-bits.

### 9.8.2.3 Square Root Computation

Applications have recently emerged in which it is important to calculate the square roots of field elements. Efficient square rooting is especially important in the case of pairing-based cryptography over binary fields (Barreto et al. 2007). Fast calculation of square roots in $F_{2^m}$ requires that all exponents in the irreducible polynomial ($a$, $b$ and $c$) be odd (Fong et al. 2004). Such irreducible polynomials are easy to find, but unfortunately most of the standard polynomials (e.g: irreducible polynomials recommended by NIST) are not of this form.

As pointed out by Fong et al. (2004), if using an irreducible polynomial with all odd exponents, for example, the trinomial $z^m + z^a + 1$, the field square root can be expressed as

$$\sqrt{a(z)} = A_{even} + (z^{(m+1)/2} + z^{(a+1)/2}).A_{odd} \qquad (9.4)$$

where $A_{even}$ and $A_{odd}$ are the even and odd indexed bits of $a(z)$ concatenated into half-sized bit arrays. Multiplication by the term $z^{(m+1)/2}$ is simple and is performed by storing $A_{odd}$ as the higher order bits of the final result. Multiplication of $A_{odd}$ by the rest of the terms in $f(z)$ can be performed with some xor operations and shifts, when necessary. The amount of calculations required for this step depends on the particular irreducible polynomial. The resulting polynomial $c(z) = \sqrt{a(z)}$ is of a degree less than $m$-1; therefore, the reduction step is not necessary.

## 9.9  Summary

Finite field arithmetic is one of the main building blocks of every IBC protocol. Finite fields are very important constructs because they have special properties that can be exploited for the purpose of cryptography. They serve as elementary blocks in ECC because elliptic curves used in cryptography are always defined over finite fields. The efficient implementation of finite field arithmetic is an important prerequisite in elliptic curve systems because curve operations are performed using arithmetic operations in the underlying field. Operations on field elements, such as addition, subtraction, multiplication, squaring, and inversion operate on numbers that are usually hundreds of bits long.

The two main types of finite fields that are suitable for implementation of elliptic curve systems are prime fields $F_p$ and binary fields $F_{2^m}$. Arithmetic in finite fields is the fundamental building block in every ECC implementation. The performance of the whole system depends in large degree on the efficiency of the arithmetic operations. The most important operation in the prime field is the multiprecision integer multiplication.

This chapter reviewed methods for calculating multiplication on low-end processors. Most of the arithmetic implementations on embedded devices are focused on prime finite fields. The choice of the field is dictated by the fact that basic arithmetic operations can be effectively optimized if pseudo-Mersenne primes are used in $F_p$.

Binary fields are not favored because binary polynomial arithmetic (multiplication, in particular) is insufficiently supported by current CPUs. The implementation results prove that arithmetic in the binary field can be efficiently implemented on different embedded processors and that it is an attractive alternative when it comes to Elliptic Curve Cryptography systems implementation. In this chapter, we reviewed the applicability of binary field arithmetic to low-end processors and examined a few methods for implementing arithmetic operations in finite binary fields.

## 9.10 References

Barreto, L. M., S. Galbraith, C. O'Heigeartaigh, and M. Scott. Efficient pairing computation on supersingular abelian varieties. *Designs, Codes and Cryptography* 42 (2007): 239–271.

Brickell, E. F., D. M. Gordon , K. S. McCurley, and D. B. Wilson. Fast exponentiation with precomputation. *Advances in Cryptology-Proceedings of EUROCRYPT 1992, Lecture Notes in Computer Science*. Springer-Verlag, 1993, 200–207.

Certicom. *Standards for efficient cryptography – SEC 2: Recommended elliptic curve domain parameters*. Certicom, 2000.

Comba, P. Exponentiation cryptosystems on the IBM PC. *IBM System* (1990): 526–538.

Eberle, H., A. Wander, N. Gura, S. Chang-Shantz, and V. Gupta. Architectural extensions for elliptic curve cryptography over GF(2m) on 8-bit microprocessors. *16th International Conference on Application-Specific Systems, Architectures and Processors*. IEEE, 2005, 343–349.

Fong, K., D. Hankerson, J. Lopez, and A. Menezes. Field inversion and point halving revisited. *IEEE Transactions on Computers* 53 (2004): 1047–1059.

Gura, N., A. Patel, A. Wander, H. Eberle, and S. C. Shantz. Comparing elliptic curve cryptography and RSA on 8-bit CPUs. *6th International Workshop on Cryptographic Hardware and Embedded Systems (Lecture Notes in Computer Science)*. Springer, 2004, 119–132.

Hankerson, D., L. J. Hernandez, and A. Menezes. Software implementation of elliptic curve cryptography over binary fields. *Cryptographic Hardware and Embedded Systems - CHES 2000: Lecture Notes in Computer Science*. Springer, 2000, 1–24.

Knuth, D. E. *The art of computer programming*. Addison-Wesley Longman, 1997.

Koyama, K., and Y. Tsuruoka. Speeding up elliptic cryptosystems by using a signed binary window method. *Advances in Cryptology; Lecture notes in Computer Science* (Springer-Verlag) 740 (1993): 345–357.

Lim, C. H., and P. J. Lee. More flexible exponentiation with precomputation. *14th Annual International Cryptology Conference on Advances in Cryptology*. Springer-Verlag, 1994, 95–107.

Liu, A., and P. Ping. TinyECC: A configurable library for elliptic curve cryptography in wireless sensor networks. *7th International Conference on Information Processing in Sensor Networks*. IEEE, 2008, 245–256.

Lopez, J., and R. Dahab. High-speed software multiplication in GF(2m) without precomputation. *INDOCRYPT: First International Conference on Progress in Cryptology*. Springer-Verlag, 2000, 203–212.

Malan, D. J., M. Welsh, and M. D. Smith. A public-key infrastructure for key distribution in TinyOS based on elliptic curve cryptography. *IEEE SECON 2004: 1st Annual IEEE Communications Society Conference on Sensor and Ad Hoc Communications and Networks* (Oct. 4–7, 2004). IEEE, 2004, 71–80. [[AU: OK AS NOW SET?]]

Miyaji, A., T. Ono, and H. Cohen. Efficient elliptic curve exponentiation. *International*

*Conference on Information and Communications Security: Lecture Notes in Computer Science.* Springer-Verlag, 1997, 282–290.

Montgomery, P. Modular multiplication without division. *Mathematics of Computation* (American Mathematical Society) 44 (1985): 519–521.

NIST. Recommended elliptic curves for Federal government use. *National Institute of Standards and Technology,* 1999. http://csrc.nist.gov/groups/ST/toolkit/documents/dss/NISTReCur.doc

Paar, C. Implementation options for finite field arithmetic for elliptic curve cryptosystems. *Invited presentation at the 3rd Workshop on Elliptic Curve Cryptography (ECC'99).* 1999.

Pohlig, S., and M. Hellman. An improved algorithm for computing logarithms over GF(p) and its cryptographic significance. Edited by IEEE. *Transactions on Information Theory* 24 (1978): 106–110.

Pollard, J. M. Monte Carlo methods for index computation (mod p). *Mathematics of Computation* (American Mathematical Society) 32 (1978): 918–924.

Seo, S. C., D. Han, and S. Hong. TinyECCK: Efficient elliptic curve cryptography implementation over G(2) on 8-bit MICAz mote. *IEICE Transactions.* IEICE, 2008.

Shantz, S. C. *From Euclid's GCD to Montegomery multiplication to the great divide.* TR-2001-95 2001, Sun Microsystems Laboratories, 2001.

Szczechowiak, P. Cryptographic key distribution in wireless sensor networks using bilinear pairings. Ph.D. thesis, Dublin City University, 2010.

Szczechowiak, P, L. B. Oliveira, M. Scott, M. Collier, and R. Dahab. NanoECC: Testing the limits of elliptic curve cryptography in sensor networks. *European Conference on Wireless Sensor Networks, Lecture Notes in Computer Science* . Springer, 2008, 305–320.

Uhsadel, L., A. Poschmann, and C. Paar. Enabling full-size public-key algorithms on 8-bit sensor nodes. *European Workshop on Security and Privacy in Ad Hoc and Sensor Networks: Lecture Notes in Computer Science.* Springer, 2007, 73–86.

Wang, H., B. Sheng, and Q. Li. Elliptic curve cryptography based access control in sensor networks. *International Journal of Security and Networks* 1 (2006): 127–137.

Woodbury, A. D., D. V. Bailey, and C. Paar. Elliptic curve cryptography on smart cards without coprocessors. *The Fourth Smart Card Research and Advanced Applications.* 2000.

Yan, H., and Z. J. Shi. Studying software implementations of elliptic curve cryptography. *3rd International Conference on Information Technology: New Generations.* IEEE, 2006, 78–83.

# Appendix—For Further Reading

Aranha, D. F., C. Gouvêa, J. López, L. B. Oliveira, M. Scott, and R. Dahab. TinyPBC (software). *Fully Enabling PKC in SensorNets.* 2010. https://sites.google.com/site/tinypbc/

Blake, I. F., G. Seroussi, and N. P. Smart. *Advances in Elliptic Curve Cryptography.* Cambridge University Press, n.d.

Chatterjee, S., and P. Sarkar. *Identity-Based Encryption.* Springer, 2011.

Gunasekaran, R., and H. Qi. XLRP: Cross Layer Routing Protocol for Wireless Sensor Networks. *Wireless Communications and Networking Conference (WCNC).* IEEE, 2008, 2135–2140.

Joye, M., and G. Neven. *Identity-Based Cryptography.* IOS Press, 2009.

Kuorilehto, M., M. Kohvakka, J. Suhonen, P. Hämäläinen, M. Hännikäinen, and T. D. Hämäläinen. *Ultra-Low Energy Wireless Sensor Networks in Practice - Theory, Realization and Deployment.* John Wiley & Sons Ltd, 2007.

Lidl, R., and H. Niederreiter. *Introduction to Finite Fields and Their Applications.* Cambridge University Press, 1986.

Martin, L. *Introduction to Identity-Based Encryption.* Artech House, 2008.

Menezes, A. J., P. C. van Oorschot, and S. A. Vanstone. *Handbook of Applied Cryptography.* CRC Press, 2001.

Muralidharan, G. *On the distributed revocation of nodes in sensor networks.* Master's thesis. College Park: University of Maryland, 2004.

Ren, K., S. Yu, W. Lou, and Y. Zhang. *IEEE Transactions on Vehicular Technology* (IEEE) 58, no. 8 (2009): 4554–4564.

Rosen, K. H., editor. *Handbook of Elliptic and Hyperelliptic Curve Cryptography.* CRC Press, 2006.

Silverman, J. H. *The Arithmetic of Elliptic Curves.* 2nd ed. Springer-Verlag, 2009.

Srivastava, V., and M. Motani. Cross-layer design: a survey and the road ahead. *Communications Magazine* (IEEE) 43, no. 12 (2005): 112–119.

Szczechowiak, P., and M. Collier. TinyIBE: Identity-based encryption for heterogeneous sensor networks. *International Conference on Intelligent Sensors, Sensor Networks and Information Processing.* IEEE, 2009, 349–354.

Yang, X., S. Xuemin, and D. Ding-Zhu. *Wireless Network Security.* Springer, 2007.

Yasmin, R., E. Ritter, and G. Wang. An authentication framework for wireless sensor networks using identity-based signatures: Implementaion and evaluation. *Institute of Electronics, Information and Communication Engineers (IEICE) Transactions on Information and Systems* (IEICE) E95-D, no. 01 (2012): 126–133.

Zheng, J., and A. Jamalipour. *Wireless Sensor Networks - A Networking Perspective.* John Wiley & Sons, Inc., 2009.

# Index